Urban Permaculture Evolution

Practical Strategies & Inspiring Visions

AF824176

Dr. D. Aly III, JD, MPA
Dr B. Charania, LLB, MD, FRCS

Copyright © 2025
All Rights Reserved
ISBN: 978-1-972004-03-6

About the Author

Dr. D. Aly III, JD, MPA

Dr. Aly is an eighth-generation landowner, investor, and sustainability advocate focused on advancing ecological innovation within modern development. As Co-Founder of Habitat Farms & Preserves, he oversees global initiatives promoting biodiversity, land restoration, and legislative reforms to prevent land fragmentation. His diversified portfolio in real estate, agriculture, and sustainable infrastructure drives the creation of self-sustaining Micro-Cities that fuse urban design with regenerative land management.

Dr B. Charania, LLB, MD, FRCS

Dr. Charania is an Indo-American physician, naturalist, and environmentalist recognized for integrating medical science with ecological health. His work emphasizes nutritional medicine and regenerative agriculture as foundational to human and planetary well-being. During his tenure with the World Health Organization and UNICEF, he contributed to the eradication of childhood infections and iodine deficiency programs worldwide. As a founding member of Charania Hospital, SIUT North America, Honorary faculty of Agha Khan Foundation, and Trustee of the Eternal Gandhi Museum, Dr. Charania's academic and humanitarian endeavors advance the nexus of public health, sustainability, and global welfare.

Preface

Cities are often seen as the antithesis of nature, concrete jungles where human ambition overshadows the rhythm of the earth. Yet, within these same bustling environments lies an extraordinary opportunity: to reimagine how we live, grow, and share resources in a way that heals rather than harms.

Urban permaculture challenges the conventional boundaries between nature and civilization. It invites us to think differently about food, energy, water, and community. It is not simply a gardening method or a set of ecological principles; it is a philosophy of living rooted in reciprocity, resilience, and regeneration.

As the world faces the combined pressures of rapid urbanization, climate instability, and social inequality, the need for systems that balance human and ecological well-being has never been greater. Permaculture provides that bridge, blending indigenous wisdom, ecological design, and modern technology into living systems that can thrive in the densest city blocks.

The chapters that follow explore the practical and visionary aspects of this transformation. From rooftop gardens to zero-waste communities, from digital mapping tools to community governance, each section reveals how urban life can be redesigned through cooperation with natural systems rather than competition against them.

This book was written for urban dreamers, architects, policy makers, educators, and anyone who believes that sustainability is not a luxury but a necessity. Whether you live in a high-rise apartment or manage an entire city district, the ideas here will help you build regenerative systems that nourish people and the planet alike.

Let this journey be a seed, one that grows into countless gardens, resilient communities, and renewed hope for the cities of tomorrow.

Introduction

The Urban Century

We are living in the *urban century*. More than half of humanity now lives in cities, and by 2050, nearly seventy percent of the global population will call urban areas home. Cities are engines of culture and innovation, yet they are also epicenters of pollution, inequality, and ecological strain.

Conventional urban planning has treated nature as an amenity, something to be contained in parks or decorative green roofs. Permaculture turns this notion upside down. It proposes that *nature is the infrastructure*, the foundation upon which thriving, resilient cities must be built.

When we apply ecological principles to urban systems, every street, rooftop, and courtyard becomes part of a living organism: cycling nutrients, capturing water, generating energy, and fostering community well-being.

What Is Urban Permaculture?

The word *permaculture* comes from "permanent agriculture" and later expanded to mean *permanent culture*. Bill Mollison and David Holmgren originally conceived it in the 1970s as an ecological design philosophy that mimics the patterns and relationships found in natural ecosystems.

Urban permaculture adapts these principles to the built environment. It focuses on **density, diversity, and interconnection**, designing systems where food production, water management, waste reuse, and social cooperation exist in harmony.

Instead of isolated projects, urban permaculture creates integrated networks of rooftop gardens that connect to community composting, which in turn feeds neighborhood farms, which supply local kitchens and schools.

Why Urban Permaculture Matters Now

- **Climate Resilience:** Cities produce 75 percent of global carbon emissions. Integrating permaculture principles reduces dependency on fossil fuels through local food systems, passive design, and renewable energy.
- **Food Security:** As global supply chains falter, urban food production ensures access to fresh, healthy produce within walking distance.
- **Water Sustainability:** Rainwater harvesting, greywater systems, and regenerative landscapes restore the hydrological cycle in the urban context.
- **Social Equity:** Community gardens and cooperative projects become social equalizers, uniting residents across economic and cultural divides.
- **Mental and Physical Health:** Green spaces heal, offering connection, exercise, and purpose in environments often defined by isolation.

Urban permaculture is not a utopian dream. It is a practical, evidence-based framework that countless communities are already adopting from Havana to Melbourne, from Nairobi to Berlin.

The Design Philosophy

At its heart, urban permaculture follows three ethics:

1. **Earth Care:** nurture the soil, water, and all living systems.
2. **People Care**: ensure human needs are met equitably and with dignity.
3. **Fair Share**: set limits to consumption and redistribute surplus for the good of all.

These ethics inform a set of design principles such as *observe and* interact, *produce no waste*, *use and value diversity*, and *integrate rather than* segregate.

In a city, that might mean transforming vacant lots into food forests, turning waste heat from buildings into energy for nearby greenhouses, or designing mixed-use developments where residents share tools, bikes, and gardens instead of duplicating them.

Permaculture design is both science and art: it analyzes flows of energy and matter, but it also celebrates creativity, culture, and beauty as essential to resilience.

Bridging Technology and Tradition

Modern cities thrive on technology, and permaculture does not reject it. Instead, it asks technology to serve life.

Digital mapping tools, smart sensors, and AI-driven irrigation systems can amplify the efficiency of ecological design. Yet, just as valuable are the low-tech practices of composting, seed saving, and communal decision-making.

The most successful urban permaculture projects fuse the two: a rooftop aquaponics garden monitored by an app, run by youth trained in traditional planting cycles.

Who This Book Is For

This book speaks to multiple audiences:

- **Urban planners and architects** seeking ecological frameworks for design.
- **Educators and activists** promoting environmental literacy.
- **Homeowners, tenants, and students** wanting to transform small spaces.
- **Policy makers** shaping the sustainable cities agenda.
- **Businesses and cooperatives** exploring regenerative economic models.

No matter your background, the message is the same: permaculture begins wherever you are.

How to Use This Book

Each chapter blends concepts, strategies, and case studies. Practical exercises and examples are included so readers can adapt ideas to their local contexts.

While you can read it sequentially, each chapter also stands alone as a reference guide.

The journey begins with understanding the core **Principles of Urban Permaculture,** the foundation upon which every subsequent system is built.

Contents

Principles of Urban Permaculture 1

Designing Urban Food Systems 9

Water Harvesting and Management 19

Renewable Energy Integration 28

Urban Housing and Architecture: Implementing Advanced Design Strategies for Dense Urban Environments 37

Designing For Urban Environments: How Urban Tree Systems, Guilds, Food Forests, And Carbon Sinks Create Climate Resilience 46

Waste Reduction: Zero-Waste Strategies, Compost Cycles, And Reuse Systems Within Communities And Cities 54

Urban Micro-Farms: Intensive Design For Small Spaces 62

Edible Landscaping for Public Spaces: Parks, Medians, and Green Corridors 72

Creating Healing Landscapes: Therapeutic Gardens and Wellness Design 89

Community Participation and Governance 98

Policy and Advocacy: Shaping City Planning with Permaculture Principles 107

Permaculture Design Certification: Pathways to Professional Practice 117

Permaculture Business Models: CSA, Cooperatives, And Social Enterprises 126

Global Case Studies: Permaculture in Cities across Continents 136

Digital Tools And Technology In Urban Permaculture Design 147

Case Studies around the World 158

The Future of Urban Permaculture: Regenerative, Resilient, and Thriving
Cities 168

Conclusion 177

Appendix 182

Principles of Urban Permaculture

Reimagining the Urban Ecosystem

In nature, there is no waste, no isolation, and no stagnation. Every element, soil, water, organism, and sunlight interacts within an endless web of exchange. This simple observation forms the philosophical core of permaculture: **design systems that mimic nature's efficiency and harmony.**

When these ideas are applied to cities, we begin to see our urban environments not as static structures, but as living ecosystems. Roads become arteries, buildings act as trees that harvest light and shade, and communities function like symbiotic species sharing resources and knowledge.

Permaculture in the city is about recognizing this latent ecology, and then consciously designing systems that strengthen it.

The Three Ethics of Urban Permaculture

Permaculture is guided by three interwoven ethics that serve as a compass for every design decision. They are both philosophical and practical, principles to live by, not just to plan with.

1. Earth Care

Earth Care means maintaining the health of the soil, air, water, and all forms of life.

In urban settings, this might translate into:

- Restoring soil health in compacted plots or abandoned lots.
- Designing green roofs that capture rainfall and provide pollinator habitats.
- Choosing materials that minimize toxicity and energy use.

Cities are often seen as the primary agents of destruction, yet they can also be the centers of restoration. Even a small courtyard planted with native species can become a microhabitat for birds, bees, and beneficial insects.

2. People Care

Sustainability is meaningless without social well-being. People Care emphasizes the creation of communities that are inclusive, cooperative, and compassionate.

In practice, this may look like:

- Designing shared gardens that bring neighbors together.
- Establishing cooperative housing where residents share tools, vehicles, and childcare.

- Encouraging participatory decision-making where everyone's voice contributes to the design.

When people are supported and connected, they naturally become better stewards of the Earth.

3. Fair Share (Return of Surplus)

The third ethic completes the cycle. Fair Share calls for responsible limits to consumption and a conscious return of surplus, whether that surplus is food, time, money, or knowledge.

In urban systems, this means:

- Sharing excess produce through community fridges.
- Repurposing waste materials for new construction or art.
- Mentoring younger generations in ecological literacy.

Fair Share ensures that the benefits of regeneration circulate equitably throughout society.

Twelve Core Design Principles

Permaculture design is built upon a set of twelve interrelated principles developed by **David Holmgren**. In an urban context, each principle offers a lens for innovation and problem-solving.

1. Observe and Interact

Every city block has its own rhythm, wind patterns, sun exposure, cultural habits, and traffic flow.

Observation is the first act of design. Before altering a space, spend time learning how it behaves. A single week of observation can save years of inefficient design.

2. Catch and Store Energy

Cities are full of wasted energy from rainwater flowing into drains to heat radiating off rooftops.

Capturing and storing energy means designing systems that retain those flows: rain barrels, solar panels, battery banks, compost bins.

3. Obtain a Yield

Every system should provide something tangible in return: food, beauty, shade, or income. In urban spaces, yield can be measured in fresh herbs from a windowsill garden or the psychological relief of greenery in a concrete block.

4. Apply Self-Regulation and Accept Feedback

Cities evolve rapidly. A design that works today might fail tomorrow.
Continuous feedback, from community input, monitoring sensors, or observation, allows systems to self-correct without collapse.

5. Use and Value Renewable Resources and Services

The most resilient systems rely on what nature freely provides: sunlight, wind, compost, and human creativity. Replacing disposable materials and fossil fuels with renewable alternatives is not just sustainable; it is economical in the long term.

6. Produce No Waste

In nature, the concept of waste does not exist. One system's output becomes another's input.

In cities, this could mean turning restaurant scraps into compost, using greywater for irrigation, or upcycling materials into furniture and art.

7. Design from Patterns to Details

Look for recurring patterns; the branching of streets resembles the branching of rivers; the clustering of homes mirrors colonies in nature.
Once the pattern is clear, fill in the details. A pattern-based approach ensures scalability and harmony within complexity.

8. Integrate Rather Than Segregate

Urban design often isolates residential areas here, commercial zones there, people divided by income or function.

Permaculture dissolves these boundaries. Mixed-use neighborhoods, shared spaces, and community hubs bring elements together for mutual benefit.

9. Use Small and Slow Solutions

Bigger is not always better. Small, low-cost interventions, such as a composting corner, a seed library, and a rooftop garden, often create bigger change than massive infrastructure projects.

Cities flourish when change grows organically, one small success at a time.

10. Use and Value Diversity

Diversity builds resilience. A monoculture city, economically, socially, or ecologically, is fragile.

Diverse plantings, cultures, and income levels create interdependence, preventing systemic collapse.

11. Use Edges and Value the Marginal

Edges, the spaces between, are where innovation happens.

In urban contexts, this might mean activating alleyways, rooftops, or abandoned lots. These transitional spaces often hold the greatest potential for transformation.

12. Creatively Use and Respond to Change

Cities never stay still. Permaculture teaches flexibility, to adapt designs as climate, technology, and culture evolve.

The ability to pivot while staying grounded in ethics ensures long-term regeneration.

Pattern Literacy: Seeing Like Nature

Patterns are the language of life. Spirals, waves, branches, and webs, these shapes repeat across scales, from the veins of leaves to the networks of roads.

Permaculture designers study patterns to understand **how energy moves**, how sunlight reaches a courtyard, how water flows off a roof, and how people circulate through a neighborhood.

For example:

- A **branching pattern** can guide stormwater management or pedestrian movement.
- A **spiral pattern** may inspire efficient garden design in tight spaces.
- A **web pattern** informs decentralized networks for food or energy distribution.

Pattern literacy transforms cities from chaotic clusters into elegant systems of flow.

The Urban Ecosystem Model

A functional urban permaculture design views the city as an **ecosystem** with the following interdependent components:

Element	Ecological Function	Urban Equivalent
Sunlight	Primary energy source	Solar panels, skylights, passive heating
Water	Circulation and nourishment	Rainwater harvesting, greywater reuse
Soil	Foundation of life	Composting, green roofs, micro-gardens
Plants	Producers, stabilizers	Urban forests, food gardens
Animals	Pollinators, recyclers	Bees, birds, compost worms
Humans	Designers, caretakers	Community networks, cooperative governance

Recognizing these parallels shifts our mindset:

The city is not separate from nature. It is nature, reorganized.

From Principle to Practice

Permaculture principles are not theoretical ideals. They are design verbs, actions that invite experimentation.

When applied to the city, they might inspire:

- **Community food forests** on vacant land.
- **Rain gardens** that filter stormwater runoff.
- **Composting networks** linking households and restaurants.
- **Energy cooperatives** producing power from solar rooftops.
- **Cultural guilds** where artists, farmers, and technologists collaborate.

These initiatives begin as local sparks but can grow into systemic transformation. The key is design rooted in ethics, scaled through participation.

The Designer's Mindset

An urban permaculture designer is both scientist and storyteller.
They observe patterns, map flows, and build prototypes, but they also cultivate imagination, empathy, and collaboration.

Cities thrive not just through technical innovation but through shared purpose.

To "design with nature" in an urban context is to remember that **ecology is not external; it is us.**

Designing Urban Food Systems

Rethinking the Source of Our Food

Food is at the heart of every civilization, yet modern cities have built themselves on a paradox: the further we advance, the further we move away from the land that sustains us. Our meals travel thousands of miles before reaching our plates. Supermarkets overflow with abundance, while most urban dwellers have never seen where their food actually grows.

Urban permaculture seeks to **close this distance** to restore the link between people, place, and nourishment. It does so not by romanticizing the rural past, but by redesigning the urban present: turning rooftops, balconies, courtyards, and abandoned lots into productive ecosystems that feed both body and community.

In a world facing climate shocks, population density, and economic uncertainty, **localized food systems** are not luxuries; they are lifelines.

Principles of Urban Food Design

When designing food systems for cities, permaculture integrates **ecology, culture, and practicality**. Each design should:

1. **Fit the context**: work within the climate, space, and social realities of the area.
2. **Close the loop**: recycle nutrients, water, and energy within the system.
3. **Build community resilience**: engage residents, schools, and local businesses.
4. **Increase biodiversity**: cultivate a range of plants and pollinators to strengthen the ecosystem.

Urban food systems thrive when they are diverse, decentralized, and community-driven, a web of micro-systems, not a single large farm.

The Layers of Urban Food Production

1. Rooftop Gardens

Flat rooftops are among the city's most underused assets. With proper design, they can become lush oases that reduce heat, absorb carbon, and grow food.

Lightweight soil mixes, drip irrigation, and wind barriers make rooftop cultivation viable in nearly any climate.

Rooftop gardens can host:

- Salad greens, herbs, and compact vegetables.
- Pollinator-friendly flowers to support bees.

- Compost bins for closed-loop nutrient cycles.

Beyond food, rooftop systems reduce building temperatures, manage storm water, and reconnect residents with the cycles of growth.

2. Balcony and Vertical Gardens

For high-rise living, **vertical gardening** brings greenery to even the smallest apartments.

Modular wall planters, hanging hydroponic towers, and recycled containers can produce herbs, strawberries, and leafy greens year-round. These micro-systems encourage urban dwellers to participate directly in the act of growing, even if their entire "garden" fits inside a single window frame.

The psychological benefits are as important as the nutritional ones: growing food reconnects people to the rhythm of life in environments that often feel mechanical.

3. Community Gardens

Few spaces capture the spirit of urban permaculture more than the community garden.

These shared plots transform vacant lots into living classrooms where neighbors learn, plant, and harvest together.

Each community garden becomes a microcosm of cooperation:

- Local composting reduces neighborhood waste.
- Rainwater catchments irrigate shared beds.

- Skill exchanges and workshops foster education and trust.

Community gardens build more than food security; they build **social security**: a network of relationships grounded in mutual care.

4. Urban Farms

Urban farms scale up production to meet local demand while staying true to ecological principles.

They may occupy rooftops, greenhouses, or per-urban land on the city's edge. Successful urban farms often combine **agroecological practices** with social enterprise, selling produce to local restaurants, farmers' markets, and CSA (Community Supported Agriculture) members.

Urban farms also serve as **educational hubs**, offering training in sustainable agriculture, composting, and small business management.

5. Aquaponics and Hydroponics

Where soil is scarce or contaminated, water-based systems offer elegant alternatives.

Hydroponics uses nutrient-rich water to grow plants directly, while **aquaponics** adds fish, creating a symbiotic cycle: fish waste feeds the plants, and the plants purify the water.

These closed-loop systems:

- Use 80–90% less water than conventional agriculture.
- Can operate indoors or on rooftops year-round.
- Provide both plant and protein yields.

Aquaponics represents the fusion of ecological wisdom and technological innovation, permaculture in its most futuristic form.

6. Food Forests

Food forests are multilayered systems modeled on natural woodlands, with trees, shrubs, vines, herbs, and ground covers all producing food.

In cities, **mini food forests** can be designed in parks, schoolyards, or along sidewalks.

A mature food forest may include:

- **Canopy layer:** fruit or nut trees (apple, fig, and almond).
- **Understory** shrubs (berries, citrus, and currants).
- **Herbaceous layer:** vegetables and herbs.
- **Ground cover:** clover, sweet potato, strawberries.
- **Root layer:** carrots, radishes, turmeric.
- **Vertical layer:** vines and climbers.

Each layer supports the others, creating self-sustaining abundance.

Over time, the forest becomes a public pantry, carbon sink, and sanctuary all in one.

Designing for Limited Space

One of the biggest misconceptions is that food production requires large areas.

Urban permaculture thrives precisely **because** it works within constraints.

Key strategies include:

- **Stacking functions:** A wall trellis provides shade, grows beans, and supports pollinators.
- **Intercropping:** Mixing fast- and slow-growing species for continuous yield.
- **Container gardening:** Using recycled barrels, crates, or buckets for mobility and flexibility.
- **Vertical layering:** Combining climbing plants with ground covers to maximize sunlight use.
- **Succession planting:** Replacing harvested crops immediately with new seedlings for year-round productivity.

The goal is to **use every cubic meter** efficiently, turning scarcity into creativity.

Closing the Loop: Nutrient and Water Cycles

No food system is sustainable without closing the loops of **nutrients and water**.

In permaculture design, the outputs of one process become the inputs of another:

- Kitchen scraps → compost → soil enrichment.
- Greywater → filtration → irrigation.
- Organic waste → biogas → cooking fuel.

Simple technologies such as **composting toilets, worm farms, and rain gardens** make these cycles tangible and accessible even in dense neighborhoods.

Social and Economic Dimensions

A regenerative food system is not just ecological, it is social and economic.

Urban agriculture provides:

- **Employment opportunities** for youth and marginalized groups.
- **Educational programs** in schools and universities.
- **Local food markets** that circulate wealth within communities.
- **Reduced dependency** on external supply chains.

Every dollar spent on local produce generates multiple returns, economic, environmental, and psychological.

By creating *food sovereignty*, cities can weather disruptions and empower citizens to feed themselves.

The Aesthetic Dimension

Food systems also reshape the **visual and emotional landscape** of cities. Edible plants soften concrete edges, improve air quality, and reintroduce beauty into everyday spaces.

A fruit-bearing street tree offers both shade and sustenance; a garden on a median strip becomes a sign of renewal rather than neglect.

When citizens see food growing in public spaces, a subtle cultural transformation occurs; people begin to believe that cities, too, can nourish life.

Case Study: The Havana Model

After the collapse of the Soviet Union in the 1990s, Cuba faced severe fuel shortages and food insecurity. Havana responded by transforming vacant lots and rooftops into **organopónicos**, urban organic gardens.

Within a decade:

- Over 35,000 hectares were cultivated inside the city.
- 90% of Havana's vegetables came from local production.
- Urban farmers became respected community leaders and educators.

Havana's story shows that urban agriculture is not just a response to crisis; it is a **blueprint for resilience** that other cities can follow.

Case Study: Detro it is Green Renaissance

Once a symbol of industrial decline, Detroit has reinvented itself through grassroots farming movements.

Organizations like **Detroit Black Community Food Security Network** and **Keep Growing Detroit** have transformed derelict lots into thriving farms.

These spaces now provide:

- Organic produce for local markets.

- Educational programs for youth.
- Models of cooperative ownership and land stewardship.

Detro it is evolution demonstrates that **abandoned land is not wasted land**; it is opportunity waiting for care and vision.

The Future of Urban Food Systems

As cities continue to expand, urban food systems will evolve beyond gardens and farms into **integrated infrastructures**:

- Smart greenhouses connected to local grids.
- AI-driven composting and irrigation networks.
- Community kitchens linked to neighborhood farms.
- Policy frameworks that mandate edible landscapes in zoning laws.

Urban food systems of the future will not just *exist within* cities, they will *define* them.

Design Exercise

Observation Task:

Take a walk in your neighborhood. Identify at least three places, a rooftop, a sidewalk edge, or a wall, where food could potentially grow.

Sketch or photograph them. Then imagine how these spaces could connect: where would compost come from? Where would water flow? Who would benefit?

This simple exercise transforms passive observation into active design, the first step in creating a regenerative city.

Water Harvesting and Management

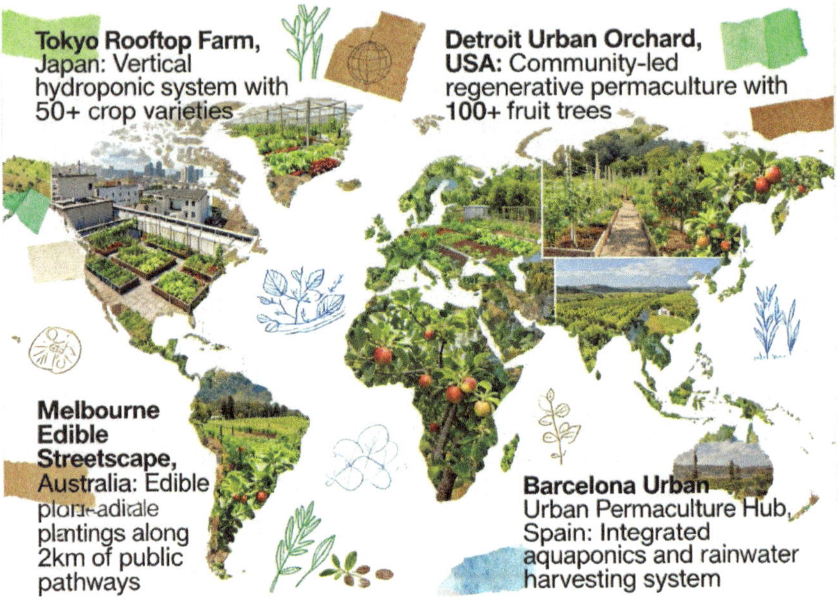

The Sacred Flow of Water

If soil is the foundation of life, then water is its pulse.

Every drop carries energy, nutrients, and memory, shaping landscapes, sustaining ecosystems, and defining the livability of cities.

Yet in most modern urban systems, water is treated as a commodity or a nuisance: piped in, consumed, and flushed away. Stormwater is diverted into drains rather than absorbed into the ground. Rivers are channelized, wetlands filled, and rain seen as something to control rather than cherish.

Urban permaculture restores the sacred relationship between **cities and the water cycle**. It recognizes that water is not just a resource, it is a living system to be integrated into design at every scale, from a balcony planter to a metropolitan watershed.

Rethinking Urban Hydrology

Traditional city planning follows a linear model: **capture → use → dispose.** Permaculture replaces this with a **circular model: harvest → store → filter → reuse**.

In a regenerative city:

- Rainwater does not flood streets; it infiltrates gardens.
- Greywater does not vanish down the drain; it irrigates trees.
- Roofs become reservoirs; wetlands become filters.

This shift, from drainage to retention, turns urban landscapes into *water catchment systems* rather than water wastelands.

Principles of Urban Water Design

Permaculture applies four simple yet profound principles to water management:

1. **Slow it**: Reduce runoff speed using vegetation, swales, and permeable surfaces.
2. **Spread it**: Distribute water evenly across the landscape.
3. **Sink it**: Allow water to infiltrate into the soil where it can recharge groundwater.
4. **Store it**: Capture excess for later use through tanks, ponds, or cisterns.

These principles ensure that every drop of rain benefits the ecosystem rather than contributing to erosion or pollution.

Water Harvesting Techniques

1. Rainwater Harvesting

Rainwater is the purest and immediate source of water for cities.

Simple systems can be installed almost anywhere:

- **Rooftop catchments** direct rain into storage tanks or barrels.
- **First-flush diverters** remove initial contaminants.
- **Filtration units** ensure water is clean enough for irrigation or even potable use with further treatment.

On a larger scale, entire apartment complexes or municipal buildings can be equipped with underground cisterns that store thousands of liters of rainwater for dry seasons.

Beyond utility, rainwater harvesting reconnects citizens with natural cycles, teaching them to *see the rain as wealth, not waste.*

2. Swales and Contour Drainage

A **swale** is a shallow, vegetated trench built along the land's contour to capture and slow runoff.

In urban areas, swales can:

- Be integrated into parks, sidewalks, and medians.
- Filter pollutants through plant roots.
- Recharge groundwater tables naturally.

Swales are both functional and aesthetic; they turn dull drainage ditches into living streams that pulse with seasonal rain.

3. Greywater Recycling

Greywater, lightly used water from sinks, showers, and laundry, makes up nearly 50% of household wastewater.

Instead of sending it to sewage systems, it can be filtered and reused for:

- Landscape irrigation.
- Toilet flushing.
- Urban agriculture.

Greywater systems can be simple (gravel and reed-bed filters) or advanced (constructed wetlands or bio filters).

In dense housing complexes, shared greywater systems dramatically reduce municipal water demand while nurturing greenery year-round.

4. Permeable Surfaces

Concrete and asphalt prevent infiltration, causing flash floods and pollution.

Replacing them with **permeable materials**, gravel, porous pavers, or green pavement, allows rain to seep into the soil below.

Sidewalks, parking lots, and plazas can all be redesigned to **absorb rather than repel** water, turning impermeable cities into sponges instead of drains.

5. Urban Ponds and Constructed Wetlands

Artificial ponds and wetlands provide multiple ecological services:

- Store excess rainwater.
- Filter pollutants through aquatic plants.
- Provide habitats for birds and amphibians.
- Moderate local temperatures.

In neighborhoods, these water features can double as community spaces, places where children play, wildlife thrives, and water once again becomes visible, tangible, and sacred.

When properly designed, wetlands act as **living kidneys** for cities, cleansing and regulating the urban ecosystem.

6. Green Roofs and Rain Gardens

Green roofs absorb rainwater, filter it through plant roots, and release it slowly into the atmosphere through evapotranspiration.

Similarly, rain gardens, shallow basins planted with deep-rooted species, capture runoff from roofs or driveways and allow it to soak into the ground.

Together, these systems reduce flooding, support pollinators, and beautify urban landscapes.

They also create microclimates that cool buildings, reduce energy use, and improve mental health by bringing nature closer to daily life.

Designing a Water Network

A truly regenerative water system operates like a **closed-loop network.**

Each stage feeds the next, ensuring nothing is wasted.

Example System Flow:

1. **Rainwater collection** from rooftops →
2. **Storage in cisterns →**
3. **Filtration through bio-sand or plants →**
4. **Distribution for irrigation or domestic use →**
5. **Greywater reuse →**
6. **Constructed wetlands for final polishing →**
7. **Infiltration into soil or aquifer recharge.**

At every stage, design elements must be placed strategically according to elevation, using gravity instead of pumps whenever possible to minimize energy demand.

Community-Scale Water Management

While individual households can do much, community-level water design multiplies impact.

Neighborhood-scale initiatives can include:

- **Shared cisterns** connected to multiple roofs.
- **Stormwater parks** that double as recreational spaces.

- **Municipal greywater systems** for irrigating public landscapes.
- **Education programs** that train residents to monitor local water quality.

When communities co-own their water infrastructure, stewardship becomes a shared responsibility rather than an outsourced service.

Case Study: Portland's Green Streets Initiative (USA)

Portland, Oregon, is a global model of urban water permaculture.

Through its **Green Streets Initiative**, the city redesigned over 1,000 intersections and sidewalks with bioswales, rain gardens, and permeable pavements.

Results:

- 94% reduction in stormwater runoff to sewers.
- Decreased flooding in vulnerable districts.
- Significant improvement in water quality entering the Willamette River.

Portland proved that reimagining streets as ecological corridors creates cities that are not only functional but also beautiful and resilient.

Case Study: Bangalore's Tank System (India)

Centuries before modern infrastructure, Bangalore developed a series of interconnected lakes and ponds, known as **tanks**, to capture and store monsoon rain.

This ancient network sustained the city for over 400 years before urbanization disrupted it.

In recent decades, local communities and NGOs have revived many of these tanks through:

- **Desilting and replanting native species.**
- **Constructed wetlands for pollution control.**
- **Community stewardship programs.**

The restored tanks now recharge aquifers, support fisheries, and reduce urban flooding, a testament to the enduring wisdom of traditional hydrology integrated with modern planning.

Integrating Technology and Traditional Wisdom

Water permaculture honors both **ancient knowledge and modern innovation.** Smart sensors, IoT-based irrigation systems, and digital flow meters can now track water use in real-time, ensuring efficiency and transparency.

Yet, traditional systems like **stepwells, cisterns, and stone check dams** remind us that simplicity often outperforms complexity in sustainability.

When the two merge, data-guided systems rooted in ecological principles, cities achieve a balance of intelligence and humility.

Design Exercise

Mapping Urban Water Flows:

- Observe where rainwater flows in your neighborhood during a storm.

- Note where it pools, where it disappears, and where it could be harvested.
- Sketch a small-scale design showing swales, tanks, or rain gardens that could redirect and reuse this water.

This exercise develops the designer's eye, teaching us to *see water not as a problem to drain away, but as a guest to welcome and guide.*

The Deeper Message: From Scarcity to Reciprocity

Water teaches us reciprocity. When we hold it, it nourishes us; when we waste it, it leaves us.

Urban permaculture transforms cities from consumers into caretakers, restoring the hydrological cycle that sustains all life.

The goal is simple yet revolutionary: a city where every drop is honored, every flow is circular, and every person understands their role as a guardian of the watershed.

Renewable Energy Integration

Energy as the Breath of the City

Every living organism breathes, taking in energy, transforming it, and releasing what is no longer needed. Cities are no different. They inhale sunlight, wind, and human activity; they exhale heat, motion, and waste. The problem is that most cities breathe **pollution instead of balance**, fueled by fossil systems that deplete rather than regenerate.

Urban permaculture envisions cities that **generate their own clean energy** and distribute it equitably, just as trees share sunlight across a forest canopy. By aligning energy systems with natural flows, we can transform cities from energy consumers into **net-positive ecosystems**, producing more power than they use while healing the environment that sustains them.

Principles of Energy Permaculture

Permaculture design approaches energy with the same ethics that govern soil and water:

1. **Earth Care**: reduce pollution and greenhouse gases.
2. **People Care**: ensure reliable and affordable access to energy for all.
3. **Fair Share**: produce surplus energy and feed it back into the grid or community.

Instead of asking, *"How much energy do we need?"* we begin by asking, *"How can we design systems that need less?"*

Energy efficiency is not merely about conservation; it is about **intelligence**, aligning design with the rhythms of nature so energy flows freely where and when it is needed most.

The Hierarchy of Renewable Design

Urban energy design follows a simple hierarchy of priorities, from **reducing demand** to **producing surplus.**

1. Conserve and Optimize

Before generating new energy, minimize what is wasted:

- Use natural light and ventilation.
- Design buildings for thermal efficiency (orientation, insulation, and shading).
- Switch to low-energy appliances and LED lighting.

- Implement behavior-based conservation (timed systems, shared utilities).

In permaculture, *the cleanest energy is the energy never consumed.*

2. Capture and Store Renewable Energy

Once waste is minimized, focus on harnessing what nature freely provides:

- Sunlight through photovoltaic panels.
- Wind through micro turbines.
- Water flow through micro-hydropower.
- Organic waste through biogas systems.

Each source becomes part of a web of flows that keeps the city alive, decentralized, modular, and resilient.

3. Share and Regenerate

Finally, distribute and reinvest surplus energy within the community:

- Feed excess solar into local micro grids.
- Trade or share stored energy between neighbors.
- Reinvest profits into expanding renewable capacity or community welfare.

This creates **energy democracy**, where ownership and benefit belong to people, not corporations.

Urban Solar Design

1. Rooftop Photovoltaics

Rooftop solar remains the most accessible renewable source in cities.

By converting sunlight directly into electricity, solar panels can power homes, schools, and offices.

Modern systems allow:

- **Net metering:** feeding excess energy back to the grid.
- **Battery storage:** ensuring resilience during outages.
- **Community solar:** shared installations for apartment dwellers or renters.

2. Building-Integrated Photovoltaics (BIPV)

Instead of panels as add-ons, **BIPV** integrates solar cells into building facades, windows, and roofs.

Transparent solar glass, solar shingles, and photovoltaic awnings transform entire structures into power plants.

This design approach redefines architecture as **living energy skin,** buildings that breathe sunlight like leaves.

3. Passive Solar Design

Permaculture emphasizes *design before technology.*

Buildings can harness the sun passively through orientation, material choice, and spatial planning:

- South-facing windows for winter warmth.
- Overhangs or shading trees for summer cooling.
- Thermal mass walls to absorb and release heat gradually.

These timeless techniques drastically reduce the need for mechanical heating or cooling, proving that elegance and efficiency can coexist.

Urban Wind and Kinetic Energy

While large wind farms may belong to rural landscapes, **small-scale wind systems** can thrive in cities with thoughtful design:

- **Vertical-axis turbines** capture multidirectional breezes and fit on rooftops.
- **Building-channeling designs** accelerate wind flow between structures.
- **Kinetic pavements** generate power from footsteps in high-traffic areas.

By blending aesthetics and function, wind trees, artistic turbines, or energy-harvesting sculptures, renewable energy becomes part of the **urban identity** rather than industrial clutter.

Biogas and Waste-to-Energy Systems

Urban waste is an untapped power source.

Organic waste from households, restaurants, and markets can be transformed through **anaerobic digestion** into:

- **Biogas:** methane-rich fuel for cooking or heating.

- **Digestate:** nutrient-rich residue for compost and soil.

Small-scale digesters fit in community gardens, food markets, or apartment complexes, turning garbage into gold while reducing landfill emissions.

Waste-to-energy models close the loop: the city feeds itself while cleaning up after itself.

Microgrids and Decentralized Power

Traditional grids are centralized, fragile, and hierarchical. Permaculture energy systems favor **microgrids**, localized networks of producers and consumers who can operate independently if needed.

Benefits of microgrids:

- Energy resilience during blackouts or disasters.
- Community ownership and local jobs.
- Flexible integration of solar, wind, and storage.

Imagine a neighborhood where each home has rooftop solar, shared batteries, and the ability to trade electricity peer-to-peer through blockchain or digital platforms.

That neighborhood does not just use energy; it **creates community through energy.**

Energy Storage and Sharing

Storage is the bridge between abundance and scarcity.

Battery systems, from lithium-ion to community-scale saltwater or gravity batteries, allow cities to store solar energy by day for use at night.

But storage isn't only mechanical. **Thermal storage** in water tanks, stalls, or even underground chambers can regulate building temperature naturally.

Permaculture teaches us to store energy **in many forms**, biological (trees, biomass), chemical (batteries), and social (knowledge and cooperation).

The true battery of a sustainable city is not silicon, its community.

Case Study: Freiburg, Germany, The Solar City

Freiburg stands as a global symbol of renewable urbanism. After the oil crisis of the 1970s, citizens resisted nuclear expansion and turned instead to solar energy.

Today:

- Over 50% of buildings generate renewable power.
- The Vauban district operates as a near-zero-emission neighborhood.
- Residents co-own energy cooperatives and reinvest profits locally.

Freiburg's model proves that grassroots activism plus municipal support equals systemic change.

The city's motto could well summarize permaculture energy design: *"Think globally, act locally, power collectively."*

Case Study: Masdar City, United Arab Emirates

Masdar City demonstrates a high-tech approach to renewable design in an extreme desert environment.

Built to operate on solar power and minimal waste, its infrastructure

Includes:

- Smart grids balance demand in real time.
- Solar concentrators and wind towers for passive cooling.
- Driverless electric transit systems.

While highly engineered, Masdar offers a glimpse of what happens when architecture, technology, and ecology merge, a laboratory for sustainable urbanism on a grand scale.

Cultural and Social Energy

Energy is not just electrons; it is emotion, movement, and connection.

Permaculture expands the concept to include **social energy**, the enthusiasm, cooperation, and creativity of people.

A community gathering in a park, a shared repair workshop, or a music event powered by solar panels all generate *human energy* that sustains more than the grid ever could.

In this sense, **culture itself becomes renewable**, powered by participation and shared purpose.

Design Exercise

Your Local Energy Map

- Observe where sunlight falls most abundantly in your building or neighborhood.
- Identify sources of wasted heat or kinetic energy (vent exhausts, moving crowds, and machinery).
- Sketch how renewable technologies, solar, wind, or biogas, could harness these flows.
- Estimate how much energy your community could generate collectively.

Designing this map reveals how much potential already surrounds us, waiting to be activated.

From Dependence to Interdependence

When cities depend on distant power plants, they externalize responsibility. When they generate and manage energy locally, they **reclaim agency**.

Renewable integration is not just an engineering challenge; it is a cultural awakening. It teaches communities to live within their ecological means while sharing abundance wisely.

Urban permaculture envisions cities that shine like living stars, each neighborhood radiant with self-sustaining energy, each rooftop a leaf in the vast canopy of a regenerative planet.

Urban Housing and Architecture: Implementing Advanced Design Strategies for Dense Urban Environments

Cities as Living Organisms

Every building, street, and courtyard in a city can be seen as a cell in a vast living organism. When these cells are healthy and interconnected, the city thrives; when they are isolated, polluted, or lifeless, the organism begins to fail.

Traditional architecture has too often focused on aesthetics or economic efficiency alone, ignoring the cycles of energy, water, and human wellbeing that sustain true livability. Urban permaculture reframes architecture as a **biological act**, where every structure participates in the regeneration of the ecosystem around it.

In this model, architecture does not dominate the landscape; it **coexists with, contributes to, and communicates** with it.

The Ethic of Regenerative Design

Regenerative design goes beyond sustainability. Sustainability seeks to minimize harm; regeneration seeks to reverse damage and actively restore life.

A regenerative building is one that:

- Produces more energy than it consumes.
- Harvests and purifies water.
- Enhances biodiversity.
- Supports human health and community interaction.
- Adapts over time to changing needs.

Such buildings do not merely "use resources efficiently." They become **producers** of oxygen, food, shade, and joy.

The Challenges of Density

High-density urban areas pose unique challenges:

- Limited space for natural ecosystems.
- High-energy demand and waste generation.
- Heat islands caused by concrete and asphalt.
- Social isolation despite physical proximity.

Yet these challenges are also opportunities. Density allows shared infrastructure, collective energy systems, and vibrant social life, if designed wisely.

Permaculture responds by **weaving nature into the vertical dimension** of the city, turning walls, roofs, and courtyards into living systems.

Principles of Permaculture Architecture

1. **Design with Nature, Not Against It**: Study the site, its sunlight, wind direction, rainfall, and noise patterns, before designing. Every building should respond uniquely to its local ecosystem.

2. **Form Follows Flow.** Instead of imposing rigid geometry, follow the natural movement of air, light, and people. Corridors become breezeways; atriums become light wells; rooftops become rain catchers.

3. **Multi-Functionality**: Every element should serve several functions. A wall can insulate, reflect light, grow plants, and collect rainwater. A courtyard can ventilate, socialize, and cultivate.

4. **Local Materials and Circular Construction:** Use what the region provides: stone, bamboo, recycled glass, earth blocks. Reuse demolition materials to reduce embodied energy.

5. **Adaptability and Longevity:** Design for change. Movable partitions, modular layouts, and flexible utilities allow buildings to evolve with community needs.

Passive Design Strategies

Passive design uses natural forces, sun, wind, water, and thermal mass, to maintain comfort with minimal energy input. It transforms buildings from mechanical consumers into natural regulators.

1. Orientation and Shading

- Position buildings to maximize winter sunlight and minimize summer heat.
- Use deciduous trees, pergolas, or adjustable louvers for dynamic shade.
- Urban canyons (narrow street corridors) can be designed to channel cooling breezes.

2. Natural Ventilation

- Cross-ventilation through operable windows and vents.
- Venturi effects using vertical shafts or wind towers.
- Courtyards that act as "lungs" for entire housing clusters.

These systems reduce dependency on air conditioning and connect occupants to natural rhythms.

3. Thermal Mass and Insulation

- Thick earthen or stonewalls absorb daytime heat and release it at night.
- Green roofs and walls insulate against temperature extremes.
- Reflective surfaces minimize urban heat buildup.

4. Daylighting

Natural light reduces electricity use and enhances mood.

Skylights, clerestory windows, and light shelves bring sunlight deep into buildings, reducing artificial lighting needs by up to 70%.

Biophilic Design: Healing Through Connection

Humans are wired to thrive in contact with nature; this is the foundation of **biophilic design.**

Integrating natural elements into urban architecture enhances both physical and psychological well-being.

Key biophilic strategies include:

- **Living walls** that clean air and reduce noise pollution.
- **Indoor gardens** and atriums that provide microclimates.
- **Natural materials** like wood and clay that regulate humidity and temperature.
- **Visual connections** to sky, greenery, and water, even if through framed views or rooftop access.

A single green wall can lower room temperatures, absorb CO_2, and calm the human nervous system. In a dense city, biophilia is not a luxury; its **urban medicine**.

Vertical Ecosystems

With limited horizontal space, permaculture embraces **vertical design**.

High-rise structures can integrate multiple layers of ecological function:

- **Rooftop farms and orchards.**
- **Hydroponic facades** for food and air purification.

- **Rainwater channels** flowing down exterior surfaces into storage tanks.
- **Solar facades** that generate electricity while shading interiors.

The "vertical forest" concept, popularized by Stefano Boeri is Bosco Verticale in Milan, demonstrates how residential towers can host hundreds of trees, thousands of plants, and entire microclimates.

Each tree becomes part of a living architectural ecosystem.

Social Architecture: Designing for Belonging

Sustainable housing is not only ecological, it is **socially regenerative.**

Isolation is one of the greatest modern diseases, and architecture often reinforces it.

Permaculture-based housing designs foster **interaction and mutual care:**

- Co-housing communities share gardens, workshops, and kitchens.
- Common courtyards encourage gatherings, markets, and play.
- Shared infrastructure like laundry, composting, and tool libraries builds interdependence.

Buildings can heal loneliness when they are designed to **invite encounters rather than enforce separation.**

Affordable and Inclusive Design

Regenerative housing must be accessible to all.

This requires innovation not just in architecture, but in economics:

- **Modular prefabrication** reduces costs and waste.
- **Local materials** minimize transportation and support nearby industries.
- **Community land trusts** prevent displacement and keep housing permanently affordable.
- **Participatory design processes** give residents a voice in shaping their spaces.

When people co-design their homes, they invest emotionally and take collective ownership of maintenance and improvement.

Case Study: Bed ZED, Beddington Zero Energy Development (London, UK)

Completed in 2002, BedZED remains a landmark in sustainable urban housing.

It integrates:

- Passive solar heating and natural ventilation.
- Renewable energy generation through solar panels and biomass.
- On-site water recycling and green roofs.
- Mixed-use spaces for living, work, and recreation.

BedZED achieved an impressive **81% reduction in energy use** compared to conventional housing, proving that ecological living can be both comfortable and beautiful.

Case Study: Hammarby Sjöstad (Stockholm, Sweden)

Originally, an industrial area, Hammarby Sjöstad was transformed into a circular urban district based on the "Hammarby Model." Its design integrates:

- Biogas production from household waste.
- Greywater purification through wetlands.
- District heating powered by renewable energy.
- Green corridors connecting housing to parks and waterways.

The neighborhood operates as a **closed-loop urban metabolism**, where outputs from one system become inputs for another, a living embodiment of permaculture principles.

Case Study: Earthship Communities (New Mexico, USA)

Earthships, developed by architect Michael Reynolds, are off-grid homes built primarily from recycled materials like tires, bottles, and earth.
They combine passive solar heating, rainwater harvesting, and indoor food production.
In dense urban adaptations, Earthship-inspired buildings demonstrate that even waste can become the foundation for comfort, autonomy, and artistry.

Design Exercise

Sketch Your Living Building:

- Draw your current living space or a favorite building.

- Identify where natural elements could be introduced, sunlight, airflow, greenery, or water.
- Imagine each surface serving multiple roles: roof as garden, wall as power source, and courtyard as cooling system.
- Ask: *How can this space give back to the Earth more than it takes?*

This exercise helps you see architecture not as shelter from nature, but as a **continuation of it**.

The Architecture of the Future

The city of the future is not a grid of lifeless towers; it is a living mosaic of ecosystems.

Buildings breathe, streets collect rain, and walls feed people.

Each resident becomes a caretaker of both their dwelling and the planet.

Urban permaculture shows that we do not need to escape cities to live sustainably; we need only to redesign them as places where life flourishes again.

In dense environments, architecture becomes poetry, a collaboration between human intention and natural intelligence.

Designing For Urban Environments: How Urban Tree Systems, Guilds, Food Forests, And Carbon Sinks Create Climate Resilience

Cities as Forests in Disguise

From above, a city looks like an artificial forest, vertical trunks of steel and glass, Asphalt Rivers, and canopies of cables. Yet unlike real forests, cities **consume more than they produce**. They emit carbon instead of absorbing it; they shed heat instead of sheltering life.

Permaculture challenges this imbalance. It asks: *What if our cities could behave like forests again?*

A forest captures sunlight, filters air, retains water, and recycles nutrients, all while providing habitat, food, and beauty. Urban permaculture applies these same ecological functions to human-made environments, turning city blocks into **living systems of regeneration**.

Trees and plants are not decoration; they are the **infrastructure of resilience**.

The Role of Urban Ecology in Climate Resilience

Cities are at the frontline of climate change. Rising temperatures, intense rainfall, flooding, and air pollution threaten both human and ecological health.

Vegetation mitigates these impacts through:

- **Carbon sequestration:** capturing CO_2 from the atmosphere.
- **Temperature regulation:** cooling surfaces and providing shade.
- **Water management:** slowing runoff and increasing infiltration.
- **Biodiversity:** creating habitats for pollinators and birds.
- **Psychological health:** reducing stress and promoting social cohesion.

By restoring ecological structure to cities, we transform them from *climate victims* into *climate healers.*

Urban Tree Systems: The Green Infrastructure

An **urban tree system** is more than scattered street trees, it is a networked canopy, planned and maintained as an ecological whole.

1. Street Canopies

Street trees lower ambient air temperatures by up to 5°C (9°F) through shading and transpiration.

They also reduce air pollution by trapping particulates and absorbing carbon dioxide.

Key design considerations:

- **Species selection:** native, drought-tolerant, and diverse.
- **Soil health:** permeable pavement and root zones that allow oxygen flow.
- **Connectivity:** trees should form continuous corridors for wildlife and cooling.

Each street becomes a **microclimate corridor**, linking parks, schools, and homes into a web of shade and life.

2. Green Roofs and Walls

Vertical greenery multiplies vegetation area in high-density zones.

Plants on roofs and facades:

- Absorb CO_2 and pollutants.
- Act as insulation, reducing energy use.
- Protect building materials from UV degradation.
- Provide refuge for insects and birds.

Even a single green wall can cool its microenvironment, absorb sound, and beautify neglected spaces.

3. Pocket Forests

Inspired by the **Miyawaki method**, pocket forests transform small urban plots into dense, fast-growing ecosystems.

Within two years, these forests can host hundreds of native trees and shrubs, creating oxygen islands that heal degraded soils and communities alike.

They demonstrate that even 100 square meters can become a thriving **biodiversity engine.**

The Guild Concept: Cooperation Among Plants

In nature, plants do not compete, they **collaborate**.
A **guild** is a group of species that work together to support a shared ecosystem function.

Example: The Urban Fruit Tree Guild

At the center stands a fruit tree, say, an apple. Around it, supporting species perform specialized roles:

- **Comfrey** mines nutrients from deep soil and provides mulch.
- **Clover** fixes nitrogen, feeding the tree.
- **Yarrow** attracts pollinators and repels pests.
- **Garlic and onions** deter insects and disease.
- **Ground covers** like strawberries protect soil moisture.

Together they create a self-sustaining micro-ecosystem, a living partnership that requires little maintenance and produces abundant yields.

In the city, guilds can be applied in planters, community gardens, or along sidewalks, providing beauty, food, and habitat simultaneously.

Food Forests in the City

A **food forest** is the ultimate expression of plant guilds, mimicking the structure of natural forests to produce food across multiple layers:

1. **Canopy:** tall fruit or nut trees (mango, oak, and walnut).
2. **Understory:** smaller trees and shrubs (citrus, berries).
3. **Herbaceous:** herbs, vegetables, flowers.
4. **Ground cover:** clover, sweet potato, creeping thyme.
5. **Root layer:** carrots, radishes, ginger.
6. **Vertical layer:** vines (grapes, beans, passionfruit).

Urban food forests can be established in parks, schoolyards, and vacant lots.

They:

- Feed communities year after year.
- Regenerate soil and absorb carbon.
- Provide shade, pollinator habitat, and education.

The forest becomes a **classroom, pantry, and carbon sink** all at once, the ultimate symbol of regenerative design.

Carbon Sinks and Climate-Positive Design

A **carbon sink** is any system that absorbs more carbon than it emits.
Urban areas, dominated by vehicles and concrete, are typically carbon sources. However, through regenerative design, they can become carbon sinks.

How Cities Can Store Carbon:

- **Tree canopies**: absorb CO_2 through photosynthesis.
- **Soil health**: compost and organic matter sequester carbon underground.
- **Biochar use**: converts waste biomass into stable carbon stored in soil.
- **Building materials**: use carbon-storing materials like hempcrete or timber.

Imagine every building facade covered with climbing plants, every sidewalk lined with soil-filled planters, and every park layered with food-producing trees.

Each element quietly, continuously **draws carbon back into the Earth**.

Designing Resilient Green Networks

Isolated green spaces cannot create resilience alone. The key lies in **connectivity**, linking ecological patches into continuous corridors that allow movement, shade, and water flow.

The Three Scales of Urban Green Networks

1. **Micro-scale:** courtyards, balconies, green roofs, and vertical gardens.
2. **Meso-scale:** neighborhood parks, green streets, and community gardens.
3. **Macro-scale:** citywide corridors connecting rivers, forests, and wetlands.

When these scales interlock, the result is a **metropolitan ecosystem**, a living infrastructure that cools, filters, and nourishes the city.

Social and Psychological Benefits

Urban greenery doesn't just heal the planet, it heals people.

Studies show that residents living near trees experience:

- Reduced stress and anxiety.
- Lower rates of cardiovascular disease.
- Improved focus and creativity.
- Stronger community bonds and reduced crime rates.

Permaculture's emphasis on cooperative maintenance, shared gardening, composting, and design, strengthens these benefits by fostering **collective care.**

When people plant and nurture trees together, they cultivate more than greenery, they cultivate belonging.

Case Study: Singapore – The Garden City

Singapore transformed from a congested port to one of the greenest urban environments on Earth.

Through integrated planning:

- Over 50% of the island is covered in vegetation.
- Every building over a certain height must include greenery.
- Sky gardens, green bridges, and vertical parks connect biodiversity corridors.
- The "City in a Garden" vision unites ecology and economy seamlessly.

Singapore's model shows that even hyper-dense cities can become **living forests** when policy, architecture, and community align.

Waste Reduction: Zero-Waste Strategies, Compost Cycles, And Reuse Systems Within Communities And Cities

The End of the Line, or the Start of a Cycle?

In nature, there is no landfill. Every fallen leaf becomes soil, every organism eventually feeds another. Waste is simply a **resource in transition**.

Human systems, however, have long broken this cycle. The linear economy, take, *make, use, discard*, has left cities drowning in their own debris. Landfills expand, oceans choke on plastic, and communities suffocate under invisible pollution.

Permaculture flips the paradigm. It treats cities as **metabolic systems**, with inputs, outputs, and feedback loops. What was once seen as garbage becomes nutrient, energy, or material for the next process.

The goal is not merely "waste management" but **waste elimination** through intelligent design. The regenerative city produces nothing that nature cannot absorb.

From Waste to Resource: The Circular Urban Economy

In a circular system:

- Products are designed to last, repair, and recycle.
- Biological materials return safely to the earth.

- Technical materials (metals, plastics, glass) are reused indefinitely.
- Energy and nutrients circulate locally rather than being lost.

This mirrors natural ecology, where each output feeds the next input, **a closed loop of life**.

A zero-waste city is not a utopia; it is simply one that has learned to think like an ecosystem.

The Three Layers of Urban Waste

Urban waste can be grouped into three main categories, each requiring unique strategies:

1. **Organic Waste**: food scraps, yard trimmings, biodegradable matter.
2. **Recyclable Waste**: metals, glass, plastics, paper.
3. **Non-Recyclable Waste**: materials that must be minimized or phased out through innovation.

Each layer offers opportunities for circular design, especially when communities participate as active stewards rather than passive consumers.

1. Composting: Nature's Recycling Factory

The Power of Decomposition

Composting is the simplest and most profound act of regeneration.

By allowing organic matter to break down in controlled conditions, we turn "waste" into rich, living soil teeming with microbes.

In the city, composting can take many forms:

- **Household compost bins:** suitable for kitchen scraps and garden waste.
- **Vermicomposting systems:** using worms to accelerate breakdown indoors.
- **Community compost hubs:** shared by multiple households, markets, or schools.
- **Municipal composting plants:** industrial-scale systems powered by organic collection programs.

The result, compost, becomes a local fertility engine for gardens, parks, and urban farms.

Composting Design Tips

- Mix "greens" (nitrogen-rich) and "browns" (carbon-rich) for balance.
- Keep moisture and aeration consistent to prevent odor.
- Use sealed bins or mesh enclosures to deter pests.
- Rotate or turn piles regularly for oxygen flow.

Proper composting can divert **30–40% of household waste** from landfills while enriching the soil that feeds the city.

2. Anaerobic Digestion and Biogas Systems

For communities producing large amounts of organic waste, **anaerobic digestion** offers a high-efficiency solution.

In sealed tanks, bacteria break down waste in the absence of oxygen, producing:

- **Biogas** (methane + carbon dioxide) used for cooking or heating.
- **Digestate**, a nutrient-rich residue usable as fertilizer.

Biogas plants can operate at neighborhood, market, or municipal scale, providing clean energy and reducing methane emissions from landfills.

This is permaculture's ethos in action: nothing wasted, everything transformed.

3. Reuse Systems: The Second Life of Materials

Before recycling comes **reuse**, extending the lifespan of materials through creativity and community.

Examples of Reuse Systems:

- **Repair cafés and fix-it workshops** where residents mend electronics, clothing, or furniture.
- **Material libraries** lending tools, books, and building supplies.
- **Upcycling projects** turning glass bottles into lamps or pallets into furniture.

- **Refill stations** for detergents, oils, and dry goods, eliminating single-use packaging.

These initiatives reduce consumption, build local skills, and create new economies of sharing, a practical manifestation of *People Care* and *Fair Share*.

4. Recycling: The Transitional Strategy

Recycling remains vital but should be seen as the **last step**, not the first. The process consumes energy and often down cycles materials (e.g., plastic into lower-grade products). Still, efficient recycling systems close important loops:

- **Paper and cardboard** → new paper or insulation.
- **Glass** → remelted into bottles indefinitely.
- **Metal** → infinitely recyclable with minimal quality loss.
- **Plastics** → require innovation toward bioplastics or modular reuse.

Cities like **San Francisco** have achieved over **80% waste diversion** through mandatory sorting, compost pickup, and recycling education, proving that collective effort drives transformation.

5. Industrial Symbiosis: The Circular City in Motion

At the city scale, waste reduction becomes **industrial ecology**, where one industry's byproduct fuels another's production.

Example: Kalundborg, Denmark

In Kalundborg's eco-industrial park:

- Excess heat from a power plant warms nearby homes.
- Wastewater from industries irrigates farms.
- Sulfur removed from flue gas is sold for fertilizer.
- Biomass waste generates biogas for energy.

This model of **industrial symbiosis** transforms the city into a self-balancing metabolism, proof that collaboration outperforms isolation.

6. Policy and Governance for Zero Waste

Cities must pair citizen action with supportive policy:

- **Extended Producer Responsibility (EPR):** manufacturers are accountable for end-of-life disposal.
- **Pay-as-you-throw systems:** households are charged based on waste volume.
- **Ban on single-use plastics and non-recyclables.**
- **Incentives for composting and material innovation.**

Government frameworks should empower community enterprises, not replace them. True zero-waste governance emerges when **policy enables participation.**

7. The Social Dimension of Waste

Zero waste is not just environmental; it is cultural. It changes how people see value and responsibility.

Community-led initiatives like **swap markets, repair fairs, and bulk-buy cooperatives** transform waste reduction into joyful participation.
Children learn composting at school. Cafés offer discounts for reusable cups. Apartment residents organize recycling drives.

When waste becomes a shared concern, it also becomes a shared opportunity for creativity and pride.

Case Study: Kamikatsu, Japan – The Zero-Waste Town

The small Japanese town of Kamikatsu achieved what many metropolises still dream of, over **80% waste diversion.**

Residents separate waste into 45 categories for recycling or reuse.

There are no garbage trucks; people take waste directly to sorting centers, learning from the process.

Repair workshops and secondhand stores turn "trash" into treasure.

Kamikatsu's motto sums up permaculture's approach:

"We don't throw things away, because there is no 'away.'"

Case Study: Curitiba, Brazil – Waste for Food and Mobility

Curitiba pioneered creative waste exchange programs in the 1990s.

Low-income residents could trade sorted recyclables for bus tickets or fresh produce from local farms.

This integrated waste reduction with social welfare, reducing landfill burden while feeding and mobilizing the poor.

It remains a brilliant example of **eco-social innovation**, using waste to nourish both ecosystems and human dignity.

Design Exercise

Your Household Waste Audit:

1. Track all your waste for one week, what you throw, compost, or recycle.
2. Categorize it: organic, recyclable, reusable, or unnecessary.
3. Identify which items could be eliminated by changing habits or suppliers.
4. Create a "personal waste reduction plan", with small but consistent actions.

You will be surprised how quickly awareness alone begins to shrink the trash bin.

From Decay to Regeneration

Zero waste is not about deprivation; it is about **abundance through intelligence**. When cities design for circularity, every discarded item becomes potential, every resident a participant in the metabolism of renewal.

The regenerative city of the future will not hide its waste; it will **celebrate its cycles**, compost piles in courtyards, refill stations on corners, art from reclaimed glass, and parks nourished by yesterday's scraps.

In such a city, nothing truly ends; it simply transforms.

Urban Micro-Farms: Intensive Design For Small Spaces

Abundance in the Tightest Corners

When people think of farming, they imagine wide fields and open skies. However, in the modern world, more people live in apartments than on farmland.

How then can we feed cities without expanding outward into forests and fields?

The answer lies not in **more land**, but in **smarter design.**

Urban permaculture teaches that productivity is not a matter of size; it is a matter of **relationship, efficiency, and diversity.**

Even a few square meters can produce surprising yields when designed as a micro-ecosystem.

Micro-farms embody the essence of urban resilience: small, adaptable, community-driven, and beautifully self-sufficient.

The Micro-Farm Philosophy

A micro-farm is more than a garden. It is a **miniature ecosystem,** intentionally designed to produce food, manage waste, and recycle energy within a confined space.

It follows the three guiding ethics of permaculture:

- **Earth Care:** build soil and biodiversity even in containers or raised beds.

- **People Care:** provide access to fresh food and nature connection.
- **Fair Share:** share surplus produce, seeds, or compost with neighbors.

Whether on a balcony, rooftop, schoolyard, or alleyway, a micro-farm redefines what is possible in the city.

Design Principles for Urban Micro-Farms

1. **Intensify through Stacking Functions** Combine vertical layers, trellised vines above, mid-level greens, and root crops below. Every cubic foot counts.
2. **Grow in Cycles, Not Rows** Replace traditional rows with guilds and polycultures, plants that benefit one another, using space and nutrients efficiently.
3. **Feed the Soil, Not the Plant** Even in containers, soil is alive. Compost teas, worm castings, and mulch sustain fertility and microbial diversity.
4. **Integrate Water and Nutrient Loops** Collect rainwater, recycle greywater, and compost on-site. Nothing leaves the system without purpose.
5. **Design for Ease and Joy** Urban farming thrives when it is simple, beautiful, and emotionally rewarding. Functionality should invite participation, not overwhelm it.

Site Selection and Preparation

Before planting, observe:

- **Sunlight exposure:** at least 4–6 hours daily for most crops.
- **Wind patterns:** strong winds dry soil; barriers or trellises can help.

- **Access to water:** rain barrels, hoses, or recycled greywater.
- **Weight limits:** crucial for rooftops or balconies, lightweight soil mixes are essential.

Prepare the site by layering organic matter, cardboard, compost, and mulch, to mimic natural forest floors. This "lasagna method" transforms even compacted urban soil into fertile ground.

Small-Space Growing Techniques

1. Raised Beds and Containers

Portable, manageable, and versatile, ideal for rooftops and courtyards.
Use recycled wood, metal troughs, or durable fabric planters.

Key advantage: control over soil composition and drainage.

2. Vertical Gardens

Walls, trellises, and shelves become productive surfaces.

Climbing crops (beans, cucumbers, tomatoes) and cascading herbs (thyme, mint, strawberries) maximize sunlight capture per square meter.

Vertical systems can also include hydroponic towers or hanging pocket planters, ideal for balconies or indoor spaces with natural light.

3. Square Foot Gardening

Divide a small area into 1-foot squares, each growing a different crop.

This method increases diversity, reduces pest pressure, and makes harvesting more manageable.

4. Companion Planting

Pairing plants that benefit one another optimizes growth:

- Tomatoes + basil (flavor and pest resistance).
- Carrots + onions (mutual pest deterrence).
- Beans + corn + squash ("Three Sisters" guild).

5. Succession and Rotation

Replace harvested crops immediately with new seedlings to maintain productivity year-round.

Rotate plant families to prevent nutrient depletion and disease buildup.

Composting in Compact Spaces

Even micro-farms must close the nutrient loop.

For limited space, try:

- **Worm bins (vermicomposting):** efficient, odor-free, and indoor-friendly.
- **Bokashi composting:** ferments food waste using beneficial microbes; ideal for apartments.
- **Mini compost drums:** aerated bins for small yards or rooftops.

The compost produced can rejuvenate soil indefinitely, a self-renewing system of fertility.

Water Management in Small Systems

Water is often the limiting factor in urban farming.

Micro-farms solve this through:

- **Drip irrigation or soaker hoses** powered by gravity.
- **Mulching** to retain moisture and moderate temperature.
- **Rainwater collection** from roofs or balconies.
- **Greywater reuse** from sinks (filtered with sand or plants).

Hydration is life, but overwatering kills as effectively as drought. The goal is balance, a steady rhythm of moist soil and living roots.

Micro-Farm Designs by Space Type

1. Balcony Micro-Farm

- Stack containers vertically along walls.
- Hang planters from railings or ceilings.
- Use light-reflective surfaces to boost photosynthesis.
- Grow herbs, salad greens, dwarf tomatoes, and edible flowers.

2. Rooftop Micro-Farm

- Use raised beds or grow bags with lightweight substrate.
- Include windbreaks (bamboo, mesh, or hedges).
- Install rain barrels for irrigation.
- Integrate pollinator-friendly plants for bees and butterflies.

3. Courtyard or Alley Farm

- Employ vertical trellises and espaliered fruit trees along walls.
- Add compost bins and shade-loving plants below.
- Use reflective walls to increase light availability.

4. Indoor Micro-Farm

- Utilize LED grow lights and hydroponic systems.
- Focus on fast-growing crops like lettuce, kale, basil, and microgreens.
- Reuse kitchen scraps for regrowth (scallions, celery, and herbs).

Micro-farms can flourish anywhere sunlight or imagination reaches.

Community and Commercial Micro-Farms

Micro-farms aren't just personal, they can scale collectively.

Community Micro-Farms

Shared plots in neighborhoods or housing complexes produce shared harvests, skill-sharing, and stronger social ties.

Youth and elders alike learn hands-on ecology, turning empty land into living classrooms.

Commercial Micro-Farms

Small entrepreneurs now use rooftops and shipping containers to produce herbs, greens, and mushrooms for restaurants and local markets.

Systems like controlled-environment agriculture (CEA) combine permaculture philosophy with modern precision, ensuring consistent yield with minimal footprint.

These ventures demonstrate that profit and sustainability can coexist when designed ethically.

Case Study: Lufa Farms, Montreal (Canada)

Built atop warehouse rooftops, **Lufa Farms** operates the world's first large-scale commercial rooftop greenhouse network.

The farms:

- Grow over 100 varieties of produce year-round.
- Use closed-loop hydroponics and rainwater capture.
- Deliver produce directly to consumers through local hubs.

Their model proves that even in cold climates, cities can become **food producers,** not importers.

Case Study: Brooklyn Grange (USA)

Brooklyn Grange manages over 5.6 acres of rooftop farmland across New York City. Each farm produces over **50,000 pounds of vegetables annually** while hosting beehives, compost operations, and educational programs. Beyond food, the farms serve as green infrastructure, cooling buildings, managing stormwater, and reconnecting citizens with nature.

Brooklyn Grange exemplifies urban agriculture as a climate strategy.

Case Study: Havana's Organopónicos (Cuba)

After economic collapse in the 1990s, Havana pioneered intensive micro-farming on vacant lots and rooftops.

Without access to fertilizers or fuel, residents turned to organic composting, crop rotation, and biological pest control.

Today, over 90% of Havana's vegetables come from within city limits.

Their model shows how a crisis can birth innovation, and how self-reliance grows resilience.

Economic and Social Impacts

Urban micro-farms:

- Provide **fresh, affordable food** in food deserts.
- Create **local jobs** in planting, distribution, and education.
- Foster community empowerment through shared ownership.
- Improve mental health and social connection.

By merging ecological and social regeneration, micro-farms become catalysts for systemic change, the smallest seeds yielding the largest transformations.

Design Exercise

Create Your Micro-Farm Blueprint:

1. Choose a space (balcony, rooftop, or shared yard).
2. Sketch the sunlight path throughout the day.
3. Design multi-level growing zones (tall plants north, short plants south).
4. Plan compost, water storage, and seating for interaction.
5. List crops by season and rotation.
6. This exercise helps you see even a 2×2 meter space as an ecosystem, where creativity, not land size, determines abundance.

From Small Seeds to Systemic Change

Urban micro-farms remind us that regeneration begins at home.

They dissolve the false barrier between rural and urban, between consumer and producer.

A single planter box can inspire a movement, a rooftop can feed a community, and a network of micro-farms can redefine a city's food security.

When thousands of small spaces act together, they form a vast invisible forest, one that breathes life into the urban fabric. This is permaculture at its purest: simple, local, beautiful, and endlessly multiplying.

Edible Landscaping for Public Spaces: Parks, Medians, and Green Corridors

Reclaiming the Edible City

In most cities, green spaces exist, but they are rarely alive in the full sense of the word. We fill parks with ornamental plants that look beautiful but feed no one, water medians lined with grass that contribute nothing to ecology, and prune trees that could just as easily bear fruit.

What if every public green space could also feed people?

Edible landscaping brings together aesthetics, ecology, and social justice. It transforms the purely decorative into the functional, making beauty productive. In doing so, it restores a fundamental truth: the landscape around us should nourish both body and spirit.

The Philosophy of Edible Landscapes

Edible landscaping is not about turning cities into farms; it is about weaving food production gracefully into public design.

The goals are:

- To create multi-layered beauty, food plants that also please the eye.
- To normalize food sovereignty, making fresh, free produce accessible to all.

- To build community interaction, shared maintenance, and harvesting.
- To strengthen ecological networks, supporting pollinators, birds, and soil life.

Layers of a Forest Garden

By blurring the boundary between garden and park, edible landscapes embody the ethics of permaculture: Earth Care, People Care, Fair Share.

Principles of Edible Public Design

1. **Visibility and Accessibility** Food should be inviting and reachable; pathways, signage, and open layouts encourage participation. A passerby should feel welcome to pick an herb, taste a berry, or water a tree.
2. **Diversity for Resilience** planting monocultures (like rows of grass or single tree species) invites pests and disease. Diversity ensures ecological health and varied yields throughout the seasons.

3. **Layering and Structure** Mimic the forest: canopy trees, shrubs, herbs, and ground covers. Each layer performs multiple roles, shading, soil building, habitat creation, and food production.

4. **Low Maintenance, High Yield** Choose perennial, drought-tolerant, and regionally adapted species to minimize labor and water use.

5. **Beauty and Function Together** Design with artistry, colors, shapes, textures, and bloom times create harmony as powerful as any ornamental garden.

Design Elements of Edible Landscapes

1. Edible Parks

Transforming existing parks into edible sanctuaries involves strategic integration:

- Replace ornamental trees with fruiting varieties (apple, fig, cherry, mulberry).
- Edge walking paths with herbs like lavender, thyme, or rosemary.
- Create demonstration gardens for school and community education.
- Include signage explaining plants' edible and ecological uses.

Edible parks invite people to reconnect with nature through taste and touch, not just sight.

2. Productive Medians and Roadside Strips

Medians, roundabouts, and roadside verges are often wasted spaces requiring costly maintenance.

By planting hardy food species and pollinator plants, these strips become green corridors of sustenance.

Examples include:

- **Dwarf fruit trees** resistant to pollution.
- **Native berries** for wildlife and humans.
- **Flowering herbs** that attract bees and improve air quality.

Such landscapes transform urban sprawl into a continuous ecological web, the city as an orchard.

3. Green Corridors and Walkways

Linear parks, bike paths, and pedestrian routes can connect neighborhoods through edible greenery.

Vines climb trellises over benches, fruit trees line shaded paths, and community gardens nestle in pocket spaces.

These corridors act as arteries of abundance, linking residents not only physically but socially.

4. Community Harvest Zones

Designating open harvest zones ensures fairness and stewardship.

Rules can be simple:

- Take only what you need.
- Help water or maintain if you can.
- Respect the plants and the people who care for them.

Such spaces empower residents to take part in nurturing the commons, shifting public land from passive consumption to shared co-creation.

Edible Plant Selection by Climate Zone

Climate	Trees & Shrubs	Herbs & Groundcovers	Perennials & Vines
Temperate	Apple, Pear, Hazelnut, Serviceberry	Sage, Thyme, Oregano, Clover	Rhubarb, Grapevine, Kiwi
Tropical/Subtropical	Mango, Guava, Papaya, Moringa	Lemongrass, Basil, Mint	Passionfruit, Sweet Potato, Taro
Arid/Semi-Arid	Olive, Pomegranate, Fig, Date Palm	Rosemary, Lavender, Purslane	Grape, Prickly Pear, Bougainvillea
Cold/Continental	Plum, Currant, Sea Buckthorn	Chives, Parsley, Comfrey	Hardy Kiwi, Hops, Strawberries

Selecting climate-adapted plants ensures sustainability and reduces maintenance, an edible landscape must thrive naturally, not struggle artificially.

Soil, Water, and Maintenance Strategies

- **Mulching and Composting:** Retains moisture, suppresses weeds, and builds soil fertility.
- **Swales and Rain Gardens:** Capture stormwater to irrigate trees and shrubs naturally.

- **Drip Irrigation:** Efficiently waters large spaces with minimal waste.
- **Community Compost Bins:** Recycle local organic waste back into public soil health.
- **Seasonal Workshops:** Teach residents pruning, harvesting, and composting, spreading skills alongside seeds.

A sustainable edible landscape runs on community participation instead of chemicals.

Case Study: Seattle's Beacon Food Forest (USA)

The Beacon Food Forest, covering over 7 acres, is one of the world's largest public edible landscapes.

Started by volunteers in 2009, it now produces fruits, nuts, vegetables, herbs, and mushrooms, all free for public harvest.

Features include:

- Terraced guild gardens following natural slope contours.
- Rainwater harvesting and composting systems.
- Educational signage and workshops for local residents.

The project has become a global inspiration for food justice, environmental education, and neighborhood solidarity.

Case Study: Incredible Edible Todmorden (UK)

In a small English town, residents turned sidewalks, schoolyards, and even police station gardens into free edible plots.

The movement, called *Incredible Edible*, now spans over 1,000 communities worldwide.

Every patch of soil, from traffic islands to churchyards, became an opportunity for growth.

Signposts read, *"Help yourself, its public food."*

The result?

- Local food security.
- Tourism and civic pride.
- Deepened community connection through shared care.

Edible landscapes, at their heart, build trust through generosity.

Case Study: The Green Pathways of Singapore

Singapore's urban planning integrates edible and ecological corridors through housing estates and commercial zones.

Public housing blocks include herb gardens, and parks are connected by tree-lined walkways with edible under plantings.

The country's *Edible Garden City* initiative has transformed rooftops and parks into community farming sites, turning limited land into limitless creativity.

Cultural Significance and Aesthetic Design

Edible landscapes can reflect cultural diversity.

Designers can incorporate plants with local heritage, indigenous species, ancestral crops, and traditional healing herbs.

A well-designed edible park might blend:

- Japanese persimmon with Mediterranean olives.
- African moringa with Andean quinoa.
- Native medicinal plants beside global culinary favorites.

This approach celebrates the multicultural fabric of urban life, making food and culture visible, tangible, and celebrated.

Governance and Policy Integration

For edible landscapes to thrive long-term, cities must:

- Include them in urban planning codes and park policies.
- Provide micro-grants or tools for community groups.
- Encourage partnerships between municipalities, schools, and NGOs.
- Adopt "Right to Grow" laws allowing public planting of edibles on verges.

When citizens and government collaborate, public spaces transform from cost centers to community assets.

Design Exercise

Mapping the Edible City:

1. Walk or map your neighborhood park or nearby public area.
2. Identify ornamental plants that could be replaced with food species.
3. Sketch where trees, herbs, fruit or vines could fit without altering public use.

4. Imagine signage or community boards inviting people to harvest and help.

This exercise cultivates a designer's eye for potential, every sidewalk, median, or verge could become a source of nourishment.

A New Civic Imagination

Edible landscapes invite a radical shift in consciousness:

From ownership to stewardship, from consumption to co-creation.

When citizens plant where they once passed by, pick where they once purchased, and share where they once competed, a new kind of city begins to bloom.

The edible city is not just a place where food grows, it is where community grows, where every hand that prunes or harvests strengthens the web of life.

To plant in public is to vote for the future.

Engaging Education, Youth Leadership, and Intergenerational Knowledge

Passing the Torch of Regeneration

Sustainability is not achieved by design alone, it is achieved by education.

A city can plant forests, build solar grids, and design zero-waste systems, but if its people do not understand why or how to care for them, those systems will wither.

Permaculture thrives only when knowledge is shared, lived, and passed on.

It grows not through manuals or mandates, but through stories, experience, and relationships between generations.

In this way, education becomes a form of cultivation, growing minds that can see patterns, think holistically, and act with compassion for life.

The Ethics of Education in Permaculture

Permaculture education is grounded in the same three ethics that guide its design:

1. **Earth Care**, teach how to protect and regenerate natural systems.
2. **People Care**; cultivate empathy, cooperation, and community resilience.
3. **Fair Share**; ensure access to knowledge for all, not just the privileged few.

Every citizen, from child to elder, is both a teacher and a student in the regenerative city.

Education as Design

Education is not just about teaching information; it is about designing environments that help people *learn through doing.*

Urban permaculture education weaves learning into daily life:

- Children learn by planting, not just reading about plants.

- Youth explore design thinking through real neighborhood projects.
- Adults learn collaboration through community gardens and repair hubs.
- Elders share traditional knowledge, seed saving, herbal medicine, and storytelling.

This integrated approach transforms the entire city into a living classroom.

Engaging Schools and Universities

1. School Gardens

School gardens are the foundation of ecological literacy.

They teach science, nutrition, responsibility, and teamwork simultaneously.

Children who grow vegetables are more likely to eat them, and to care for the land that grows them.

Key features:

- Composting systems, that close loops on food waste.
- Rainwater catchments and solar learning stations.
- Rotating student teams for maintenance and harvesting.

These spaces cultivate more than crops, they grow curiosity, confidence, and respect for natural cycles.

2. Curriculum Integration

Permaculture principles can be embedded across disciplines:

- **Science:** ecosystems, climate, energy, and soil health.
- **Mathematics:** measurement, design geometry, and data tracking.
- **Art:** botanical illustration, mural making, and pattern observation.
- **Social Studies:** local food systems, community mapping, and social equity.

Education becomes holistic, connecting intellectual learning with sensory experience.

3. University Partnerships

Colleges and universities can host **urban permaculture labs**, experimental spaces for students and researchers to test new models of food production, water management, or renewable energy.

Collaborations between academia, local government, and community organizations allow cities to become living testbeds for sustainability. Knowledge flows both ways, from research into practice, and from community wisdom into curriculum.

Empowering Youth Leadership

The next generation will inherit both the challenges and the tools of regeneration. Youth are not only learners, they are designers, innovators, and activists with the passion and creativity to reshape systems.

Pathways to Empowerment

- **Youth-led gardens and markets:** teaching entrepreneurship and teamwork.
- **Permaculture youth summits and design camps:** blending theory with practical projects.
- **Digital platforms:** connecting young change makers globally through storytelling and open-source design.
- **Mentorship programs:** pairing youth with experienced farmers, architects, or environmentalists.

Young people learn best when trusted with responsibility, when adults step back and **let them lead.**

The Power of Intergenerational Knowledge

Modern cities often separate generations, children in schools, adults in offices, elders in care facilities. This fragmentation weakens the transmission of wisdom.

In contrast, permaculture design encourages **intergenerational collaboration**, where each age group contributes distinct strengths:

- **Elders** offer memory, patience, and traditional techniques.
- **Adults** provide structure, resources, and mentorship.
- **Youth** bring innovation, energy, and digital fluency.
- **Children** remind everyone of joy, curiosity, and wonder.

Together they form a **social ecosystem**, mirroring the diversity and balance of natural systems.

Learning through Community Practice

Community Gardens as Classrooms

A community garden is not just for food, it is a place where people learn about ecology, cooperation, and conflict resolution. Workshops on composting, herbal medicine, or seed saving invite residents to share their expertise.

Education becomes communal and informal, learning **by sharing, not lecturing.**

Public Demonstration Sites

Parks, rooftops, and libraries can host demonstration plots showing principles like greywater reuse or vertical gardening.

These visible examples spark curiosity and normalize ecological living for everyone passing by.

Festivals and Celebrations

Seasonal festivals, harvest fairs, seed swaps, solstice gatherings, connect cultural traditions with sustainability.

They make permaculture not just a discipline, **but a way of life woven into community identity.**

Technology and the Digital Classroom

The digital revolution can amplify permaculture education:

- **Online mapping tools** let students analyze ecosystems and design gardens virtually.

- **Augmented reality (AR)** apps visualize green infrastructure on real city streets.
- **Online networks** connect educators and practitioners across continents.

Technology, when guided by ecological ethics, becomes a bridge, not a barrier, between knowledge and action.

Case Study: The Permaculture Institute of Los Angeles (PILA, USA)

PILA collaborates with schools and community centers to bring hands-on permaculture education into the urban core.

Programs include:

- Youth workshops on soil building and composting.
- Internships in rooftop gardening and food forest management.
- Mentorship for young designers and environmental justice activists.

Graduates often go on to lead community initiatives, creating a ripple effect across neighborhoods. It is education that grows leaders, not just learners.

Case Study: Cuba's Urban Agriculture Education Movement

After Cuba's economic crisis in the 1990s, agricultural education became central to national resilience.

Schools integrated urban farming into daily learning.

Students grew food for lunch programs, studied soil biology, and maintained compost systems.

Today, Cuban youth carry a strong sense of ecological identity, proof that **teaching self-reliance creates self-sustaining societies.**

Case Study: The "Elder Gardens" of Melbourne, Australia

In Melbourne, retirement communities have begun collaborating with local schools to maintain shared permaculture gardens.

Elders mentor students in composting, pruning, and seed saving, while learning digital design tools from younger volunteers.

These spaces dissolve generational barriers and foster mutual respect.

The result: not just gardens, but **relationships that regenerate culture.**

Design Exercise

Your Learning Ecosystem:

1. Map all learning spaces in your community, schools, parks, gardens, cultural centers, and homes.
2. Identify how they could connect to form a "learning network."
3. Ask: where can youth lead? Where can elders teach? Where can everyone meet?
4. Sketch programs or rituals that connect these nodes, compost workshops, seed festivals, storytelling nights.

Education is not confined to classrooms; it lives wherever curiosity meets care.

Education as Continuity

Permaculture education ensures continuity, the passing of torch and trowel alike. It reminds us that regeneration is not a project, but a lineage.

When a child plants a tree beside an elder, something timeless happens:
knowledge takes root, not in memory, but in action.

A truly regenerative city educates not to produce workers, but to nurture **caretakers of the Earth.**

Creating Healing Landscapes: Therapeutic Gardens and Wellness Design

The City as a Place of Healing

Modern cities are built for movement and productivity, but rarely for rest. We have engineered efficiency at the cost of stillness, speed at the cost of reflection. Amid concrete and noise, people long for places that soothe the senses and restore connection with something greater than themselves.

Healing landscapes invite us back to that connection.

They remind us that health is not only a personal matter, it is ecological. When the soil is sick, we are sick. When the water is polluted, our bodies carry that pain. When air is unbreathable, our spirits grow weary.

Urban permaculture sees healing not as a luxury, but as **a design responsibility**, the creation of spaces where people and nature recover together.

What Is a Healing Landscape?

A healing landscape is any space, garden, park, courtyard, or rooftop, intentionally designed to **restore the mind, body, and spirit**. It integrates ecology, psychology, and sensory experience.

Such spaces:

- Regulate temperature, air, and water naturally.
- Encourage relaxation and mindfulness.
- Support physical rehabilitation or gentle activity.
- Foster emotional balance through beauty and belonging.
- Reconnect individuals to the rhythms of life beyond human pace.

Healing landscapes are not "gardens for the sick", they are **gardens that prevent sickness.**

The Science of Healing Through Nature

Research consistently confirms what our ancestors always knew: contact with nature heals.

1. Physiological Benefits

- Reduced cortisol levels and blood pressure.

- Improved immune response.
- Faster recovery from surgery or illness when patients have green views.
- Enhanced sleep quality and physical mobility.

2. Psychological Benefits

- Lower stress and anxiety.
- Reduced symptoms of depression and trauma.
- Increased focus, creativity, and cognitive function.
- Social and Spiritual Benefits
- Stronger community connection through shared gardening or outdoor activities.
- Renewed sense of meaning and gratitude.
- Restoration of empathy, for others and for the planet.

When cities integrate healing landscapes, they create **public health ecosystems** that complement hospitals, not compete with them.

Principles of Therapeutic Landscape Design

Multisensory Experience

Healing happens through all senses, sight, sound, touch, smell, and even taste.

- Visual: natural colors, flowing forms, sunlight, and shade.
- Auditory: water features, wind in leaves, birdsong.

- Olfactory: aromatic herbs and blossoms.
- Tactile: textured paths, stones, bark, and soil.
- Gustatory: edible plants that invite interaction.

1. **Accessibility for All** Paths should be level and wide enough for wheelchairs and walkers. Seating should be frequent and shaded. Sensory gardens can guide visually impaired visitors through fragrance and texture.
2. **Safety and Comfort** Gentle lighting, smooth surfaces, and secure boundaries create a sense of peace and belonging. Healing requires comfort, not caution.
3. **Symbolism and Meaning** Water for renewal, circular paths for continuity, trees for strength, flowers for rebirth, symbolic elements communicate healing intuitively.
4. **Participation and Care** People heal faster when they care for something living. Community involvement in planting and maintenance builds empowerment, purpose, and shared ownership.

Types of Healing Landscapes

1. Hospital and Healthcare Gardens

Hospitals are often sterile and stressful. Integrating green courtyards and therapy gardens transforms them into **places of hope**.

Features include:

- Raised planters for patients with limited mobility.
- Shaded rest zones for families and staff.
- Fragrant plants like lavender and jasmine to calm anxiety.

- Flowing water for auditory relief.

Such gardens shorten hospital stays and reduce medication needs, a measurable return on compassion.

2. Community Healing Gardens

These are neighborhood sanctuaries where people come to heal collectively, through planting, sharing, and connecting.

They often emerge after crises: a burned-out lot, a flood site, a memorial space. Gardening becomes therapy; soil becomes medicine.

Every trowel in the earth becomes an act of grief and renewal, healing **the land and the heart simultaneously.**

3. Sensory and Meditation Gardens

Designed for mindfulness and introspection, these spaces focus on sensory harmony and simplicity.

They may feature:

- Circular walking paths for meditation.
- Soft textures underfoot, sand, moss, or woodchips.
- Soundscapes of trickling water or chimes.
- Minimal visual clutter, emphasizing light and shadow.

Such spaces restore the nervous system in a way no digital screen ever could.

4. Therapeutic Urban Farms

Urban farms can double as healing environments, especially for trauma survivors, elders, and children.

Tending soil and nurturing plants helps rebuild a sense of agency and trust. Programs that combine horticultural therapy with community farming show significant improvements in emotional resilience and physical health.

5. Green Healing Corridors

Linking hospitals, schools, and residential areas through **tree-lined walkways** or **biodiverse greenways** turns the city itself into a continuous therapy garden.

These corridors promote active mobility, clean air, and subtle daily exposure to natural rhythms.

Healing Through Design Elements

Water

Water symbolizes purification and flow. Even small features, a fountain, pond, or rain chain, evoke calm and continuity.

Moving water also oxygenates air and cools microclimates, serving both ecological and psychological purposes.

Plants

Plants with soft textures, soothing scents, and seasonal variation provide sensory comfort. Examples:

- **Lavender, rosemary, mint:** calm and refresh the mind.

- **Chamomile, calendula:** promote relaxation and healing.
- **Bamboo or grasses:** create gentle, rhythmic movement in the wind.

Pathways

Curved, looping paths slow movement and encourage mindfulness. Materials like stone, gravel, or wood allow tactile variation underfoot.

Shelter and Shade

Arbors, pergolas, or tree canopies protect visitors from harsh sun or rain, symbolizing safety

.Shade trees can lower local temperatures by up to 10°C, turning oppressive heat into comfort.

Art and Symbolic Forms

Sculptures, mosaics, and community art installations celebrate resilience and transformation.

They invite emotional expression and collective storytelling, art as therapy, rooted in place.

Case Study: Maggie's Centres (UK and Worldwide)

Founded by Maggie Keswick Jencks, these cancer care centers integrate architecture, landscape, and therapy.

Each center surrounds patients with light, gardens, and calm materials.

Instead of sterile hospitals, they offer warm, nurturing spaces, where healing is as much emotional as physical.

The success of Maggie's Centres shows that design itself can be medicine.

Case Study: The Healing Forest of Seoul, South Korea

The Seoul Healing Forest was created to help citizens recover from urban stress and trauma.

Programs include guided forest walks, breathing exercises, and horticultural therapy.

Participants report improved mood, sleep, and community connection, reaffirming that contact with nature remains one of humanity's oldest prescriptions.

Case Study: Chicago's Lurie Garden (USA)

Located in the heart of downtown Chicago, the Lurie Garden embodies the principle of resilience.

Native prairie plants absorb stormwater, attract pollinators, and bloom through all seasons.

Visitors experience calm and grounding amid skyscrapers, a metaphor for endurance in the face of modern chaos.

Design Exercise

Design Your Healing Space:

1. Identify a space, a courtyard, a balcony, or even an office window.

2. Ask: How can this place support recovery, reflection, or peace?

3. Add three sensory elements, perhaps water sound, scent from herbs, or natural textures.

4. Create a seating or meditative area that encourages stillness.

5. Reflect on how caring for this space changes your own emotional rhythm.

Healing design begins wherever care is intentional.

The Spiritual Dimension of Healing Landscapes

Healing landscapes do more than restore the body, they restore meaning.
In tending to nature, we remember that we are nature.

To heal the Earth is to heal ourselves, and to heal ourselves is to heal the Earth.

Permaculture reframes wellness as a **reciprocal process:** every garden we create to calm the human heart also cools the city, cleans the air, and feeds other beings.

The regenerative city, therefore, is not just sustainable; it is **therapeutic**. Every park, garden, and green roof becomes a living pulse of renewal, reminding us that health is not found in isolation, but in connection.

Community Participation and Governance

The City as a Commons

Architects or policymakers alone cannot build a truly regenerative city. It must be **co-created**, shaped by the people who live within it, guided by their collective values, and sustained by their shared care.

For centuries, cities have been governed by centralized systems that treat citizens as consumers rather than participants. Decisions flow downward; feedback rarely flows up. Permaculture governance reverses that flow.

It views the city as a **living commons**, a shared ecosystem of land, infrastructure, culture, and responsibility.

When everyone becomes a caretaker rather than a client, the city itself begins to heal.

The Ethics of Participation

Permaculture's three core ethics, Earth *Care, People Care, Fair Share*, extend naturally into governance:

1. **Earth Care → Ecological Policy** Laws and systems must protect soil, water, and biodiversity as foundational rights.
2. **People Care → Inclusive Decision-Making** every voice, especially those historically marginalized, must be heard in shaping the city.

3. **Fair Share → Equitable Access** Resources and opportunities are distributed justly, no community is left behind.

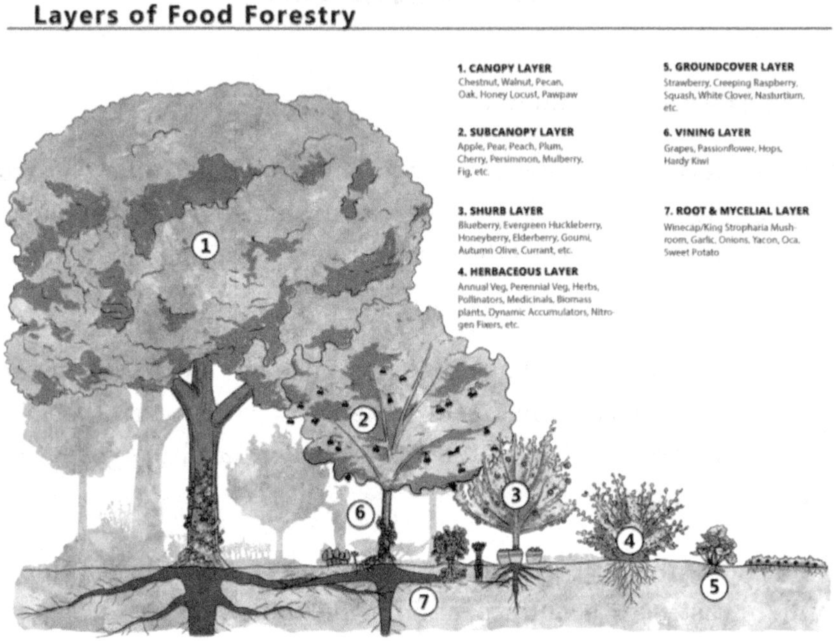

Participation is not merely political; it is **ethical design in action.**

From Top-Down Control to Bottom-Up Collaboration

Centralized urban planning often creates efficiency at the expense of humanity. Permaculture governance favors **distributed leadership**, where authority and responsibility are shared among citizens, neighborhoods, and local institutions.

This shift transforms residents from passive beneficiaries into active designers of their environment.

1. Co-Design

Residents collaborate with planners, architects, and ecologists from the very beginning of a project.

Workshops, town halls, and mapping sessions allow everyone to voice needs and propose solutions.

2. Co-Management

Communities manage public spaces together, parks, gardens, composting sites, and shared infrastructure. This builds accountability through care rather than enforcement.

3. Co-Ownership

When people own part of what they maintain, a cooperative housing unit, an energy grid, or a shared farm, they invest emotionally and financially in its success.

The Social Infrastructure of Regeneration

Physical design matters, but social design determines whether systems endure.

Strong communities share four elements:

1. **Trust**: the glue of cooperation.
2. **Transparency**: open communication and access to information.
3. **Training**: skills in facilitation, mediation, and decision-making.
4. **Tradition**: rituals and gatherings that reinforce belonging.

These create the invisible networks that hold the city together, the **mycelium of human culture.**

Models of Participatory Governance

1. Community Assemblies

Neighborhood assemblies allow residents to deliberate and vote on local initiatives.

Examples: street tree planning, food forest locations, waste management priorities.

Assemblies function best when decisions are binding, not symbolic.

2. Participatory Budgeting

Originating in Porto Alegre, Brazil, participatory budgeting lets citizens directly allocate portions of the municipal budget. This fosters transparency, trust, and innovation, communities fund projects that matter most to them.

3. Cooperative Structures

Housing co-ops, worker-owned businesses, and energy cooperatives distribute profits and responsibilities equitably.

They embody permaculture's "Fair Share" ethic in economic form.

4. Local Councils and Transition Towns

Transition Town movements create local resilience committees focusing on food, energy, and social inclusion.

They connect grassroots action to municipal policy through collaboration rather than confrontation.

Permaculture Decision-Making Tools

1. Consensus Building

All voices are valued; decisions emerge from dialogue rather than majority rule.

Consensus encourages listening, empathy, and compromise, the foundations of community harmony.

2. Sociocracy

A modern governance model based on feedback loops.

Communities are divided into "circles," each responsible for specific domains (food, energy, education).

Circles are connected through representatives, ensuring inclusivity and adaptability.

3. The Community Mandala

A permaculture-inspired tool mapping stakeholders, needs, and feedback channels.

By visualizing social relationships, communities identify strengths, gaps, and collaboration opportunities.

Reclaiming the Commons

Throughout history, shared resources, water, land, forests, have been privatized, leading to inequality and ecological degradation.

Permaculture governance seeks to **restore the commons** through co-management and community stewardship.

Examples:

- **Community land trusts** securing land for affordable housing and gardens.
- **Collective ownership** of renewable energy microgrids.
- **Open-source data and designs** shared freely for public innovation.

The commons is not a relic of the past, it is the future of equitable cities.

Conflict as Compost

No community is free from conflict, but in permaculture, conflict is **compost**, not contamination.

When managed well, disagreement produces growth and understanding.

Techniques for healthy community dialogue:

- **Nonviolent communication (NVC):** focus on needs, not blame.
- **Restorative circles:** repair relationships rather than punish.
- **Facilitated reflection:** learn from mistakes without shame.

Just as soil thrives on decay, communities thrive when differences are digested into wisdom.

Case Study: Porto Alegre, Brazil – The Birthplace of Participatory Budgeting

In 1989, Porto Alegre launched one of the world's first participatory budgeting systems. Thousands of residents met annually to set funding priorities for housing, sanitation, and transport.

Results:

- Over 20% of the city budget allocated by citizens.
- Improved infrastructure in low-income areas.
- Stronger civic engagement and transparency.

The process inspired over 2,000 cities worldwide, proof that democracy deepens when people design it together.

Case Study: The Cooperative City of Bologna, Italy

Bologna pioneered the **"Regulation on the Urban Commons"**, a legal framework that allows citizens to propose, manage, and co-govern public assets.

Over 450 pacts have been signed between residents and the city, transforming parks, schools, and vacant buildings into shared community spaces.

Bologna shows that **policy can nurture participation** when it treats citizens as partners, not problems.

Case Study: The Transition Network (Global)

Founded in Totnes, England, the **Transition Movement** empowers communities to create local resilience through permaculture-inspired projects.

Transition groups operate in over 50 countries, focusing on:

- Local food production.
- Renewable energy cooperatives.
- Skill-sharing and repair cafés.
- Emotional resilience and community storytelling.

Their success lies not in perfection but in participation, small, consistent actions multiplied across thousands of neighborhoods.

Design Exercise

Mapping the Commons:

1. Choose a neighborhood or community you know well.
2. List all shared spaces: parks, schools, community centers, sidewalks, libraries, even neglected lots.
3. Identify who uses them, who manages them, and who is excluded.
4. Envision how these spaces could be co-managed or redesigned for mutual benefit.
5. Create a "commons map" showing potential partnerships and responsibilities.

This exercise helps visualize the shift from *ownership* to *stewardship*, the cornerstone of regenerative governance.

The Culture of Participation

Community participation is not a task; it is a culture, built on daily acts of cooperation:

- Sharing tools with neighbors.
- Planting a shared garden.
- Attending a local assembly.
- Listening with curiosity, not certainty.

When participation becomes habitual, governance transforms from bureaucracy to belonging.

People stop waiting for permission, they start creating together.

From Authority to Stewardship

The regenerative city does not centralize power; it **circulates it.** Leadership becomes facilitation. Decision-making becomes dialogue. Responsibility becomes shared joy.

When every citizen is a co-designer, and every space is a commons, governance evolves from control to **care**, the highest form of leadership.

In such a city, democracy is no longer an institution; it is an ecosystem.

Policy and Advocacy: Shaping City Planning with Permaculture Principles

Designing Policy as We Design Landscapes

Just as permaculture designs ecosystems by observing patterns and relationships, **policy design** must do the same.

Cities are living organisms, systems of energy, people, and infrastructure. Yet most policies treat them as machines: mechanical, linear, and controllable.

To build regenerative cities, we must rewrite the rules of governance through the lens of ecology.

Policy becomes not a static set of laws, but a **living framework** that adapts, evolves, and restores balance, just like nature itself.

Why Policy Matters in Permaculture

Individual gardens, rooftop farms, and community initiatives can transform lives, but without supportive policies, their impact remains limited. Local laws determine whether citizens can harvest rainwater, plant fruit trees on sidewalks, or run composting systems in shared spaces.

Policy is the invisible soil in which every sustainable project must root. When it is fertile, innovation flourishes; when it is toxic, even the best intentions wither.

The Role of Advocacy

Advocacy bridges the gap between grassroots vision and institutional action. It transforms personal passion into collective power.

Through advocacy, citizens influence how cities spend resources, regulate land, and define progress.

Permaculture advocacy asks governments to recognize:

- **Ecological health as infrastructure.**
- **Community participation as governance.**
- **Regeneration as the new metric of success.**

It is not protest for protest's sake; it is the **art of alignment** between natural systems and human systems.

Principles of Permaculture-Informed Policy

1. Integrate, Don't Segregate

Traditional urban policy divides land into rigid zones, residential, industrial, and commercial. Permaculture calls for **mixed-use design,** where housing, food production, and commerce coexist in harmony.

Policy Example:

Allow rooftop agriculture, home-based composting, and small-scale food businesses in residential zones.

2. Small and Slow Solutions

Instead of grand, expensive megaprojects, prioritize local, adaptive interventions.

A decentralized network of small solutions is more resilient than a single large one.

Policy Example:

Grant micro-funding to neighborhood-level green infrastructure projects, rain gardens, micro-farms, and energy co-ops.

3. Catch and Store Energy

Policies should value renewable systems and reward conservation.
Encourage energy autonomy at every scale.

Policy Example: Feed-in tariffs for household solar, tax incentives for battery storage, or subsidies for energy-efficient retrofits.

4. Use and Value Diversity

Cities thrive on social and ecological diversity.

Policies must protect both, affordable housing, immigrant entrepreneurship, and biodiversity corridors are all expressions of resilience.

Policy Example:

Mandate native planting in public spaces and support multicultural markets as urban food nodes.

5. Produce No Waste

Waste policy must transition from disposal to circularity.

Policy Example:

Ban single-use plastics, enforce producer responsibility, and integrate composting into municipal waste systems.

6. Observe and Interact

Good policy starts with **listening**, to citizens, ecosystems, and feedback data.Pilot programs should precede permanent regulations, allowing for iterative learning.

Policy Example:

Run experimental zoning districts for regenerative projects, and evolve policy based on outcomes. Policy Domains for Urban Permaculture

1. Urban Agriculture and Food Security

Cities must treat food production as a public good.

- Legalize edible landscaping on public land.
- Simplify permitting for farmers' markets and community gardens.
- Provide land-use incentives for vertical farms and hydroponic startups.
- Include food access metrics in city resilience plans.

2. Water Management

Reform outdated drainage systems into **water-harvesting networks**.

- Require rainwater capture in new buildings.
- Promote greywater reuse in residential areas.
- Restore wetlands as stormwater infrastructure.
- Create community-led watershed councils to monitor quality and flow.

3. Renewable Energy

Shift from centralized utilities to **distributed generation**.

- Allow neighborhoods to form renewable energy cooperatives.
- Incentivize passive solar building design.
- Mandate energy audits before building permits.

Energy independence equals resilience, both ecological and political.

4. Waste and Resource Recovery

Move from waste management to **resource metabolism**.

- Mandate organic waste diversion.
- Support urban compost enterprises.
- Develop materials recovery facilities powered by renewable energy.

Cities like San Francisco and Kamikatsu demonstrate that policy-driven circular economies are achievable and profitable.

5. Housing and Land Use

Equitable access to land is the foundation of ecological justice.

- Encourage **community land trusts** to prevent gentrification.
- Integrate green roofs, vertical gardens, and shared courtyards into building codes.
- Offer tax reductions for developers who include public green amenities.
- Prioritize mixed-income eco-housing developments.

Land is not a commodity, it is a **living trust between generations**.

6. Transportation and Mobility

Transportation policy shapes how we inhabit cities.

- Expand walkable, bikeable, and tree-lined routes.
- Invest in electric public transit.

- Convert abandoned parking lots into green plazas and micro-farms.

Movement should regenerate, not deplete.

7. Health and Wellbeing

Policies that link ecology with public health reduce medical costs and increase life satisfaction.

- Integrate "green prescriptions" in healthcare systems, time in gardens, not just pills.
- Require minimum green space per capita.
- Incentivize therapeutic garden design in hospitals and workplaces.

Institutionalizing Permaculture Principles

Governments can embed ecological thinking directly into **city charters, building codes, and budgeting systems.**

For example:

- **Amsterdam's "Doughnut Economics" Model** links policy to ecological boundaries and social equity.
- **Copenhagen's Carbon-Neutral Plan** integrates energy, transport, and urban farming policies into a single resilience framework.
- **Quito's Food Charter** ensures local food systems are part of the city's disaster preparedness strategy.

These cities show that when policy mirrors ecosystem design, sustainability becomes governance, not an afterthought.

The Role of Advocacy Networks

Advocacy thrives in alliances.

Groups like:

- **ICLEI (Local Governments for Sustainability)**
- **C40 Cities Climate Leadership Group**
- **Transition Network**
- **Urban Food Policy Pacts**

...connect local movements to global influence. They provide frameworks, mentorship, and data to strengthen local advocacy efforts.

Collaboration multiplies momentum, when cities learn from each other, they evolve faster than any single policy could achieve.

Case Study: Curitiba, Brazil – Integrated Urban Ecology

Curitiba's success stems from policy coherence.

Every urban plan integrates ecology:

- Bus corridors reduce emissions.
- Parks double as flood zones.
- Recyclables are exchanged for food.
- Education campaigns make sustainability a civic identity.

It is a city designed by permaculture logic, systems feeding systems, efficiency serving equity.

Case Study: Portland, Oregon – Green Planning Framework

Portland's urban growth boundary prevents sprawl and protects farmland. Policies encourage mixed-use zoning, urban food forests, and renewable energy microgrids. Citizen input is mandatory in all major projects.

The result: one of North America's most livable, climate-resilient cities.

Case Study: Amsterdam, Netherlands – The Circular City

Amsterdam's Circular Strategy 2020–2025 applies "doughnut economics" to every policy area, ensuring that urban growth stays within planetary limits while meeting human needs.

From construction waste reuse to community composting, every policy aligns with circular economy principles.

This demonstrates how policy becomes design, and design becomes governance.

Design Exercise

Draft a Regenerative Policy Framework:

1. Choose a single urban issue, waste, water, food, or housing.
2. Identify the stakeholders: citizens, businesses, institutions.
3. Apply permaculture principles (observe, integrate, catch energy, share surplus).
4. Draft three policy recommendations that would create a closed loop.

5. Consider feedback: how would this policy evolve through real-world experience?

This practice reveals that good policy is not written, it is grown.

From Activism to Policy Design

Activism plants the seed of change, but policy ensures it takes root. The transition from protest to proposal marks the maturity of a movement.
Urban permaculture provides the language of design that policymakers can understand, systems, efficiency, cycles, and feedback.

When ecological ethics enter city planning, the walls between government and community dissolve.

Policy becomes not control from above, but **care expressed through structure.**

The Future of Governance

In the regenerative city, laws are not barriers but bridges, connecting what people need with what ecosystems can provide. Budgets reflect bioregions, not borders. Metrics measure biodiversity, not just GDP.

Citizens no longer demand sustainability, they **live it,** supported by systems that make the ethical easy.

True advocacy is not rebellion; it is redesigning civilization to work like nature again.

Permaculture Design Certification: Pathways to Professional Practice

From Philosophy to Profession

Permaculture began as a grassroots movement, a design philosophy practiced by gardeners, farmers, and environmental dreamers. But as the climate crisis deepens and urban challenges multiply, permaculture has evolved into a recognized discipline, one that blends ecology, design, and social innovation.

The Permaculture Design Certificate (PDC) is the bridge between personal commitment and professional impact.

It equips learners to design, teach, consult, and implement regenerative systems across urban and rural settings alike.

In this chapter, we explore what certification means, how to obtain it, and how it can empower practitioners to turn ideals into livelihood.

What Is a Permaculture Design Certificate (PDC)?

The PDC is an internationally recognized credential that signifies proficiency in ecological design based on the ethics and principles established by Bill Mollison and David Holmgren. It is not an academic degree, but rather a hands-on qualification rooted in experience and application.

Most courses follow a **72-hour curriculum** covering:

- Ethics and principles of permaculture.
- Design methodology and pattern understanding.
- Soil, water, and energy systems.
- Urban and rural food production.
- Community governance and economics.
- Social permaculture and teaching methods.

Upon completion, participants receive certification to practice permaculture design professionally or pursue advanced teaching credentials.

Why Certification Matters

1. Legitimacy and Professional Recognition

Certification gives practitioners credibility with clients, organizations, and governments. It signals both technical skill and ethical alignment.

2. Global Community

Graduates join a worldwide network of designers, educators, architects, and activists. This community shares knowledge, resources, and opportunities across borders.

3. Career Diversification

PDC holders apply permaculture across many fields:

- Urban planning and architecture.
- Landscape design and ecological restoration.
- Renewable energy consulting.
- Food systems and social entrepreneurship.
- Education and NGO development.

The PDC opens pathways not just to employment, but to **purposeful work.**

Core Competencies of a Certified Designer

A professional permaculture designer combines **science, art, and empathy**.

Key competencies include:

1. **Systemic Thinking**: seeing the interconnections between energy, people, and ecosystems.
2. **Observation and Analysis**: reading landscapes and communities before intervening.
3. **Integrated Design**: merging food, water, shelter, and waste systems into harmony.

4. **Communication and Facilitation**: guiding groups through collaborative processes.

5. **Ethical Practice**: applying the principles of Earth Care, People Care, and Fair Share in every project.

Certification ensures not uniformity, but **integrity**, a shared language of design that can be adapted to any context.

The Structure of a PDC Course

While **formats** vary, most PDCs share a similar framework:

1. **Foundations and Ethics:** Understanding the philosophical roots and ethical pillars of permaculture.
2. **Design Process:** Learning tools like site mapping, zoning, sector analysis, and pattern recognition.
3. **Ecosystem Literacy:** Studying soil ecology, hydrology, plant guilds, and microclimates.
4. **Urban and Social Systems:** Exploring city design, cooperative economics, and social regeneration.
5. **Hands-On Practicum:** Students apply theory to real-world projects, designing gardens, rooftop systems, or community plans.
6. **Design Presentation:** The course culminates in a group or individual project presented to instructors and peers.

Upon completion, graduates receive the **Permaculture Design Certificate**, recognized globally by institutes and networks.

Advanced Certifications and Specializations

After earning a PDC, practitioners can specialize further:

- **Permaculture Teacher Training (PTT):** prepares graduates to teach PDC courses and workshops.
- **Advanced Design Certification:** focuses on complex ecological and urban planning projects.
- **Diploma in Applied Permaculture Design:** a multi-year mentorship demonstrating practical experience.
- **Social Permaculture and Community Facilitation:** applying design principles to human networks and governance.

Each specialization deepens the designer's capacity to **translate ecological wisdom into systemic change.**

Accredited Institutions and Networks

Prominent global organizations offering recognized certifications include:

- **Permaculture Institute (USA)**
- **Permaculture Research Institute (Australia)**
- **Gaia University (International)**
- **Permaculture Association (UK)**
- **Urban Permaculture Institute (San Francisco)**
- **Regenerative Design Institute (California)**

Local hubs and grassroots collectives also host PDCs, often integrating indigenous knowledge and local bioregional focus.

What matters most is the **lineage of teaching** that instructors trace back to the founding curriculum and practice its ethics authentically.

Online and Hybrid Learning

In recent years, digital learning platforms have expanded access to PDCs worldwide. Online courses offer flexibility, while field-based sessions provide hands-on practice.

Hybrid models combine the best of both:

- Online theory modules.
- In-person design workshops.
- Mentorship through video consultations and community networks.

Virtual education has democratized permaculture, allowing urban residents, professionals, and students to gain certification without leaving their bioregions.

Ethics in Professional Practice

Certification brings responsibility. A permaculture designer must act not only as a consultant but as a **caretaker of integrity**.

Core ethical standards include:

- **Transparency** in process and pricing.
- **Community inclusion**, designing *with*, not *for*, clients.
- **Local sourcing** of materials and labor.

- **Long-term accountability**, returning to assess system health and impact.

Professional practice is an act of service, not extraction.

The designer's ultimate client is not the human, it is the ecosystem itself.

Integrating PDC Graduates into City Systems

Cities increasingly recognize certified permaculturists as essential assets in sustainability planning.

They contribute to:

- **Municipal greening programs.**
- **Community resilience initiatives.**
- **Urban agriculture policy.**
- **Corporate sustainability strategies.**

Imagine every city department, transport, housing, health, having a permaculture designer on staff.

That is the next frontier of professional practice.

Case Study: The Urban Permaculture Institute of San Francisco (UPISF)

UPISF offers city-focused PDCs integrating food systems, housing, and public space design.

Graduates have gone on to:

- Design rooftop farms for restaurants.

- Consult on city composting programs.
- Lead neighborhood food forest initiatives.

Their model demonstrates how certification translates into civic transformation.

Case Study: Gaia University (Global)

Gaia University takes an innovative approach to higher education through "learning by doing." Students design self-directed projects in regenerative systems, earning diplomas in applied permaculture, ecovillage design, or transition leadership.

This decentralized model honors diverse knowledge systems and turns **education into action.**

Design Exercise

Mapping Your Professional Path:

1. List your current skills, experiences, and interests.
2. Identify how they intersect with permaculture ethics and principles.
3. Choose one specialization (urban design, food systems, education, etc.).
4. Sketch a 3-year plan, training, projects, and partnerships to pursue.
5. Visualize how your work can serve both people and the planet.

This reflection transforms learning into livelihood, passion into profession.

The Future of Regenerative Professions

Permaculture certification is not just about gardening or design, it is about shaping the next generation of regenerative professionals:

- Architects who design living buildings.
- Economists who measure value through well-being.
- Teachers who turn classrooms into ecosystems.
- Policymakers who think like watersheds.

The PDC marks the beginning of a lifelong practice: the art of aligning human systems with natural intelligence.

When enough certified designers step into leadership, in schools, governments, and businesses, cities will evolve from extraction to regeneration.

This is more than career development; it is civilizational renewal.

Permaculture Business Models: CSA, Cooperatives, And Social Enterprises

From Profit to Purpose

In the conventional economy, profit is the goal and people or planet are the cost.

In a regenerative economy, profit becomes the byproduct of purpose, emerging naturally from systems that enhance life instead of depleting it.

Permaculture business models bridge this shift.

They are designed not only to sustain livelihoods but also to regenerate ecosystems, strengthen communities, and circulate value fairly.

By redefining success through the ethics of *Earth Care, People Care, and Fair Share*, these enterprises prove that doing good and doing well are not opposites, they are partners in resilience.

The Regenerative Economy

A regenerative economy mimics the logic of natural ecosystems:

- Waste from one process becomes input for another.
- Diversity ensures adaptability.
- Energy flows efficiently through cooperative relationships.

- Growth is cyclical, not exponential, aligned with carrying capacity.

Instead of extraction, it emphasizes circulation.

Instead of ownership, it promotes stewardship.

Instead of competition, it cultivates cooperation.

In this model, business becomes a living organism that feeds the social and ecological systems it depends on.

Core Principles of Regenerative Business Design

1. **Closed-Loop Systems:** Design products, services, and operations so all outputs become inputs elsewhere.

2. **Local Resilience:** Source locally, hire locally, and reinvest profits in community regeneration.

3. **Transparency and Accountability:** Share data, costs, and decisions openly with stakeholders.

4. **Social Equity:** Prioritize fair wages, shared ownership, and diversity in leadership.

5. **Ethical Scaling:** Grow in depth (impact), not just breadth (size).
6. **Cultural Regeneration:** Respect local traditions and indigenous knowledge as part of economic design.

These principles turn businesses into **agents of ecological literacy and cultural renewal**.

Community Supported Agriculture (CSA): Food with a Face

The CSA Model

A CSA is a partnership between farmers and consumers.

Members pay in advance for a share of the harvest, sharing both risk and reward with the producer.

This eliminates intermediaries, secures farmers' income, and connects consumers directly to the source of their food.

Benefits

- **For Farmers:** Predictable income, reduced waste, and community support.
- **For Consumers:** Fresh, local produce and personal connection to growers.
- **For the Planet:** Reduced transportation emissions, packaging, and chemical dependency.

CSAs exemplify permaculture economics, **relationship-based, seasonal, and reciprocal.**

Urban CSAs

In cities, rooftop farms, school gardens, and community greenhouses can run mini-CSA programs.

Residents subscribe for weekly produce baskets, herbs, or honey jars. Restaurants and cafes can also join as institutional members, ensuring steady markets for urban growers.

This transforms fragmented rooftops and vacant lots into a distributed local food network, nourishing both people and economy.

Cooperatives: Shared Ownership, Shared Prosperity

What Is a Cooperative?

A cooperative (co-op) is a member-owned organization where profits and decisions are shared democratically.

Each member, worker, consumer, or producer, has one vote, regardless of financial stake.

Types of Co-Ops

1. **Worker Cooperatives:** Employees own and manage the business collectively.
2. **Consumer Cooperatives:** Customers own shares and guide company decisions.
3. **Producer Cooperatives:** Small producers pool resources for marketing, distribution, or processing.
4. **Housing Cooperatives:** Residents co-own and co-manage their living spaces.

Cooperatives embody permaculture's *Fair Share* ethic, redistributing power and wealth through participation.

Benefits of the Cooperative Model

- Builds community trust and shared responsibility.
- Keeps wealth circulating locally.
- Reduces exploitation and wage inequality.
- Encourages long-term decision-making over short-term profit.

A cooperative is not just a business; it is a **social ecosystem** where everyone is both contributor and beneficiary.

Social Enterprises: The Hybrid Model

Between traditional business and charity lies the **social enterprise**, a self-sustaining venture that reinvests profits into a social or environmental mission.

Examples:

- A recycling company funding youth training.
- A café supporting food waste recovery and composting.
- A permaculture design firm reinvesting in urban greening projects.

Unlike NGOs dependent on grants, social enterprises achieve independence through entrepreneurship, while maintaining ethical accountability through purpose-driven governance.

Designing a Permaculture Enterprise

1. Start with Purpose, Not Product

Ask: *What ecological or social problem does this business heal?*

Purpose defines structure; structure determines success.

2. Apply Systems Thinking

Analyze flows of energy, materials, and relationships.

Design feedback loops so no resource leaves the system unused.

3. Create Multiple Yields

Every action should produce more than one benefit, food + education, profit + biodiversity, product + community connection.

4. Build Diversity

Mix revenue streams: membership fees, workshops, local sales, and grants.

Diversity stabilizes both ecosystems and economies.

5. Use Ethical Marketing

Communicate transparently and focus on education.

Customers are not targets; they are **participants in transformation.**

Case Study: Mondragón Cooperative (Spain)

Founded in the Basque region in 1956, Mondragón is one of the world's largest cooperative networks.

It includes over 90 worker-owned enterprises in manufacturing, finance, and education.

Core features:

- Democratic governance: one worker, one vote.
- Internal solidarity fund for struggling co-ops.
- Continuous reinvestment in training and innovation.

Mondragón proves that **large-scale cooperative economics** can be globally competitive and socially regenerative.

Case Study: Evergreen Cooperatives (Cleveland, USA)

Evergreen operates a network of worker-owned businesses providing services to local hospitals and universities, from laundry and solar installations to urban farming.

Each co-op creates local jobs, builds skills, and anchors wealth within the community.

This model demonstrates how **institutional partnerships** can fund regenerative economies in historically underserved neighborhoods.

Case Study: The Edible Schoolyard Project (California, USA)

Founded by chef Alice Waters, the Edible Schoolyard blends education, food, and community enterprise.

Students grow, cook, and sell produce, reinvesting profits into their own learning environments.

It is a living example of a social enterprise that regenerates both soil and spirit.

The Economics of Reciprocity

Permaculture business models challenge the scarcity mindset.

They operate through reciprocity, giving generates abundance.

When communities share resources, labor, and profits, wealth multiplies across the system.

This is not utopian idealism; it is ecological realism.

In nature, nothing thrives in isolation, everything depends on exchange.

Funding and Financial Resilience

- **Crowdfunding and Community Shares:** Let citizens invest directly in local projects.
- **Impact Investing:** Attract ethical investors seeking social and environmental returns.
- **Grant Partnerships:** Align with public institutions supporting green innovation.
- **Barter Networks and Local Currencies:** Encourage local trade loops that retain value within communities.

The healthiest enterprises diversify income streams just as healthy ecosystems diversify species.

Design Exercise

Design Your Regenerative Enterprise:

1. Identify a local problem, food waste, unemployment, urban heat, or disconnection.

2. Define your business mission as a healing response.

3. Sketch your inputs, outputs, and feedback loops.

4. Determine ownership: cooperative, CSA, or social enterprise.

5. Visualize how profits circulate to sustain both people and planet.

This process turns idealism into a blueprint, ethics into an enterprise.

Measuring Success beyond Money

Regenerative businesses track multiple yields:

- Soil health and biodiversity.

- Carbon sequestration and energy savings.

- Employment and education opportunities.

- Social inclusion and mental well-being.

They use triple-bottom-line accounting, people, planet, and profit, but often add a fourth: purpose.

The Future of Regenerative Enterprise

As cities transition toward ecological economies, permaculture entrepreneurs will play a vital role:

- Advising on sustainable food systems.
- Designing community-owned renewable energy.
- Rebuilding local manufacturing through circular design.
- Teaching ecological literacy in schools and institutions.

The regenerative entrepreneur is both visionary and gardener, cultivating systems that sustain life at every level.

When thousands of such enterprises bloom, the urban economy itself transforms into a living forest, rooted in ethics, branching through cooperation, and bearing the fruits of equity.

Global Case Studies: Permaculture in Cities across Continents

Learning from Living Laboratories

Permaculture is not theory, it is a global experiment. From the dense urban blocks of Tokyo to the informal settlements of Nairobi, cities around the world are rediscovering how to live within ecological limits while enhancing human well-being.

Each region faces unique cultural, climatic, and economic realities, yet all share a common thread: the desire to regenerate.

These case studies illustrate that urban permaculture is not one model to be copied, but a pattern to be adapted, proving that regeneration is universal, though its expression is always local.

1. Havana, Cuba – Food Security through Urban Agriculture

Background

In the early 1990s, Cuba faced a severe food and fuel crisis following the collapse of the Soviet Union.

Imports of fertilizers, fuel, and machinery vanished almost overnight. The country responded not with panic, but with innovation.

Transformation

Vacant lots, rooftops, and courtyards became organopónicos, organic urban gardens that rely on compost, biological pest control, and manual labor.

By 2000, Havana had over 200 urban farms producing more than 60% of the city's fresh vegetables.

Key Permaculture Elements

- Closed-loop compost systems.
- Raised-bed cultivation using local materials.
- Cooperative labor and local markets.
- Community education integrated with production.

Impact

Havana's experience proved that food sovereignty can flourish in cities, even under resource scarcity.

Today, the city's urban agriculture network serves as a global model for resilience and community empowerment.

2. Melbourne, Australia – The Urban Permaculture Capital

Background

Australia is the birthplace of permaculture, and Melbourne has become a showcase of urban adaptation.

Initiatives

- **CERES Community Environment Park:** a 10-acre sustainability hub featuring gardens, composting, renewable energy, and education programs.
- **Urban food forests** integrated into public parks and schoolyards.
- **Co-housing communities** using shared gardens and greywater systems.

Key Lessons

- Policy integration with community projects ensures continuity.
- Local government support can amplify grassroots momentum.
- Social permaculture, cooperation, communication, and inclusion, is as vital as soil fertility.

Melbourne demonstrates that permaculture can thrive in a modern metropolis when citizens and institutions collaborate.

3. Nairobi, Kenya – Regeneration from the Ground Up

Background

Nairobi faces rapid urbanization, poverty, and ecological degradation. Yet in its informal settlements, permaculture has emerged as a grassroots movement of empowerment.

Initiatives

- **The Kibera Green Movement** teaches residents to build vertical sack gardens, compost pits, and rainwater harvesting systems.
- **The Green Belt Movement,** founded by Wangari Maathai, has planted over 50 million trees nationwide, turning reforestation into civic action.

Impact

Urban permaculture has improved food security, generated income, and created climate-resilient communities.

In a place once defined by scarcity, people now cultivate abundance, proving that regeneration is not wealth-dependent, but **will-dependent**.

4. Berlin, Germany – The Green Infrastructure Revolution

Background

Berlin has transformed its post-industrial spaces into thriving green ecosystems through a culture of experimentation.

Key Examples

- **Prinzessinnengarten:** a mobile urban farm on reclaimed land, producing vegetables in movable containers.
- **Tempelhofer Feld:** a former airport turned public commons with gardens, bike paths, and pollinator corridors.
- **Circular construction policies** promoting reuse of materials from demolished buildings.

Key Permaculture Themes

- Adaptive reuse of space.
- Urban biodiversity and pollinator support.
- Participatory governance, citizens co-manage public land.

Berlin's story shows how creativity, community, and policy can turn concrete into carbon sinks.

5. Vancouver, Canada – The Eco-District Model

Background

Vancouver's "Greenest City Action Plan" integrates permaculture ethics into urban development.

Key Practices

- Eco-districts combining green buildings, renewable energy, and community gardens.
- City-wide composting and zero-waste strategies.
- Rooftop gardens and pollinator corridors spanning across neighborhoods.

Outcome

6. Bangkok, Thailand – Water, Culture, and Resilience

Background

Bangkok's low elevation and seasonal monsoons make it one of the most flood-prone capitals in the world.

Permaculture Response

- Floating gardens and aquatic crop systems (mimicking ancient Thai agriculture).
- Wetland parks designed to absorb floodwater while supporting biodiversity.
- Community-based aquaponics combining fish and vegetable production.

Impact

The integration of traditional water wisdom with modern design has created an amphibious city model, resilient, adaptive, and harmonious with its environment.

7. Portland, Oregon (USA) – The City That Gardens

Background

Portland has long embraced ecological planning and community activism.

Key Initiatives

- **Depave Project:** converting asphalt lots into community gardens and rain-absorbing landscapes.
- **City Repair Project:** residents redesign intersections with murals, benches, and micro-gardens.
- **Green Streets Program:** bioswales and rain gardens integrated into public infrastructure.

Outcome

Portland's citizens practice "civic permaculture", reshaping their neighborhoods from the bottom up through creativity and care.

The city has become a living experiment in participatory design.

8. Singapore – The Vertical Garden Nation

Background

With limited land and high density, Singapore embraced "biophilic urbanism" as national policy.

Key Initiatives

- **Gardens by the Bay:** integrates ecosystem education with spectacular design.
- **Green roofs and vertical forests** required in new developments.
- **Edible Garden City movement:** transforming rooftops and schools into productive farms.

Impact

Singapore demonstrates how advanced technology and ecological principles can coexist, creating a metropolis where nature is infrastructure, not ornament.

9. São Paulo, Brazil – The Urban Agroforestry Revolution

Background

São Paulo's marginalized peripheries have become fertile grounds for urban agroforestry, blending permaculture with social inclusion.

Projects

- **Horta das Corujas (Garden of the Owls):** a community-managed edible park open to all.
- **City policies supporting agroforestry cooperatives** and training youth in regenerative agriculture.

Results

- Restored soils in degraded areas.
- Strengthened community ties through shared harvests.
- Employment opportunities in sustainable food systems.

São Paulo's model shows how **social justice and ecological restoration** can grow together.

10. Paris, France – Agriculture on the Rooftops

Background

In response to climate targets and food insecurity, Paris launched **"Les Parisculteurs"**, a plan to cover rooftops and unused land with urban farms.

Highlights

- Over 100 new green roofs and terraces producing fruits, vegetables, and herbs.
- Integration of compost, beehives, and rainwater systems.
- Public-private partnerships blending business innovation with community access.

Lesson

Policy alignment and design creativity can transform one of the world's densest cities into a productive ecosystem, a city that feeds itself, both materially and spiritually.

Patterns Across Continents

Across all these diverse cities, clear patterns emerge:

- **Local adaptation:** success comes from designing with, not against, place.
- **Community participation:** regeneration grows through cooperation.
- **Policy alignment:** legal and institutional support amplifies impact.
- **Education and culture:** long-term change depends on shared understanding.

Urban permaculture is not confined by borders, it is a **global language of care**, spoken through soil, water, and community.

Design Exercise

Map Your City's Potential:

1. Choose your city or town.
2. Identify underused or degraded spaces, rooftops, alleys, schoolyards, or abandoned lots.
3. Research what other cities have done in similar climates.
4. Envision how one pilot project could catalyze wider transformation.
5. Connect with existing networks, you are not starting alone.

This exercise reveals that regeneration is not imported; it is **grown from context**.

The Global Movement

Permaculture in cities is not a niche, it is a movement spanning continents and cultures. it is powered by the belief that sustainability isn't about sacrifice, but about **rediscovering abundance through design.**

Each case study proves that transformation is possible, from the humblest courtyard to the grandest metropolis, when people apply creativity, ethics, and collaboration.

The next frontier of urban civilization is not technological dominance, but **ecological intelligence**, cities that function as forests, communities that govern like ecosystems, and economies that flow like rivers.

Regeneration is not the future, it is already happening.

Digital Tools And Technology In Urban Permaculture Design

Merging Ancient Wisdom with Modern Innovation

Permaculture was born from observation, the patient study of how nature designs itself. But as cities grow denser and challenges more complex, technology has become a powerful ally in **scaling ecological intelligence.**

Digital tools, from mapping software to artificial intelligence, help designers visualize ecosystems, track data, and optimize regenerative systems.

They do not replace intuition or ethics; rather, they **amplify our ability to see patterns**, connect stakeholders, and manage living systems in real time.

When guided by permaculture ethics, technology becomes not an instrument of control, but an **extension of care.**

The Digital Permaculturist

The new generation of permaculture designers blends soil literacy with digital literacy.

They understand that regeneration in cities requires both:

- hands in the soil,
- And eyes on the data.

This dual approach allows urban systems, energy, food, water, and waste, to be designed as **living networks**, constantly adapting through feedback and innovation.

1. GIS Mapping and Spatial Analysis

Geographic Information Systems (GIS) are among the most powerful tools in modern permaculture.

They allow designers to analyze and visualize spatial data: elevation, soil type, sun exposure, rainfall, vegetation, and human infrastructure.

Applications

- Mapping water flow and identifying ideal catchment or swale locations.
- Analyzing urban heat islands to target tree planting.
- Designing zoning maps for food forests, compost sites, and pollinator corridors.

- Overlaying social data (income, density, accessibility) with ecological factors to design for equity.

Common Tools

- **QGIS:** open-source, widely used for land and city-scale analysis.
- **ArcGIS:** professional software with 3D modeling and advanced analytics.
- **Google Earth Pro:** useful for beginners and visual presentations.

GIS enables designers to integrate data into holistic, evidence-based systems, transforming intuition into measurable design.

2. Drone and Aerial Technology

Drones provide rapid, high-resolution imaging of urban and peri-urban sites.

Uses

- Monitoring tree canopy cover and biodiversity.
- Tracking erosion, flooding, and heat stress.
- Creating detailed topographic maps for design planning.
- Documenting the evolution of regenerative projects over time.

When combined with GIS, drones create **digital twins** of ecosystems, allowing designers to test scenarios before implementation.

3. 3D Modeling and Simulation Tools

Three-dimensional visualization bridges the gap between concept and reality.

Popular Tools

- **SketchUp & Rhino:** ideal for architectural and landscape models.
- **Lumion or Twinmotion:** simulate lighting, vegetation, and weather impacts.
- **Revit & BIM (Building Information Modeling):** integrate sustainability data directly into construction plans.

Designers can simulate the shading patterns of trees, energy flow in buildings, or water catchment during rainfall, creating **eco-architectural harmony** between form and function.

4. Data Sensors and Smart Systems

Smart cities often use IoT (Internet of Things) technologies to manage infrastructure.

When repurposed for permaculture, these tools enhance ecological feedback systems.

Examples

- **Soil moisture sensors** trigger irrigation only when needed.
- **Air quality sensors** track pollution hotspots for planting strategies.
- **Composting sensors** monitor temperature and humidity for efficiency.

- **Water flow sensors** prevent waste and detect leaks early.

These technologies allow communities to **quantify care**, measuring regeneration in real time.

5. Renewable Energy Monitoring Platforms

Digital dashboards can now track solar output, energy consumption, and storage performance.

Apps such as Enphase Enlighten, SolarEdge, or Home Assistant integrate household or community microgrid data.

When visualized publicly (for example, on a community center wall), energy data becomes a tool for education and engagement, proving that sustainability is not abstract, but **visible and empowering**.

6. Digital Collaboration Platforms

Urban permaculture thrives on collaboration. Tools that connect citizens, designers, and local governments accelerate participation.

Useful Platforms

- **Trello / Notion / Asana:** manage community projects and task flows.
- **Miro or FigJam:** visualize design concepts in real time.
- **Slack / Discord:** foster continuous dialogue among members.
- **OpenCollective / Patreon:** fund community gardens or regenerative initiatives transparently.

Digital collaboration democratizes design, ensuring every voice has a place in shaping urban ecology.

7. Open-Source Mapping and Data Sharing

Open data transforms isolated projects into a network of collective intelligence.

Examples

- **OpenStreetMap (OSM):** allows local mapping of gardens, green roofs, compost hubs, and water sources.
- **iNaturalist:** enables citizens to record and share biodiversity data.
- **WorldClim & NASA EarthData:** offer free global datasets for climate modeling.

By merging open data with local observation, communities build living atlases of urban regeneration, continuously updated by the people themselves.

8. Artificial Intelligence and Predictive Modeling

AI tools analyze vast datasets to reveal hidden patterns, an invaluable resource for city-scale permaculture.

Applications

- Predicting **flood or drought risk** based on climate models.
- Optimizing **crop rotation** or planting cycles for rooftop farms.
- Analyzing **soil health data** to recommend compost ratios.

- Detecting **urban heat trends** and suggesting tree planting zones.

However, the key is ethical application: AI must serve life, not replace it.

Used wisely, it helps designers make decisions grounded in both data and empathy.

9. Digital Storytelling and Community Education

Technology can also cultivate emotional connection, through visual storytelling that inspires action.

Examples

- Drone videos showcasing local food forests.
- Interactive maps of community gardens.
- Short documentaries on regenerative neighborhoods.
- Online workshops or AR/VR tours of permaculture sites.

By translating ecological complexity into visual narratives, designers reach wider audiences and ignite civic imagination.

10. Blockchain and Decentralized Governance

Although often associated with finance, blockchain technology can support transparent, community-led governance.

Potential Uses

- Tokenized systems for local trade and resource exchange.

- Transparent tracking of community investments or green bonds.
- Decentralized decision-making (DAO) models for neighborhood projects.

When integrated ethically, these systems enhance accountability and autonomy, the digital equivalent of social permaculture.

Bridging the Digital Divide

While technology accelerates progress, it also risks deepening inequality.

Permaculture ethics demand inclusivity, ensuring digital tools remain accessible to all.

Strategies include:

- Offering open-source or low-cost alternatives.
- Training communities in basic mapping and digital literacy.
- Designing offline-friendly interfaces for rural or low-connectivity areas.
- Pairing tech experts with local ecological educators for co-learning.

Technology must empower, not exclude.

The most regenerative systems combine high-tech precision with low-tech wisdom.

Case Study: Smart Forest City – Cancún, Mexico

Designed by Stefano Boeri, Smart Forest City is envisioned as a carbon-neutral urban ecosystem housing 130,000 residents.

The city integrates digital monitoring systems for water, energy, and biodiversity, powered by solar panels and surrounded by millions of trees.

It demonstrates how smart technologies can serve ecological intelligence when aligned with permaculture ethics of balance and feedback.

Case Study: OpenAg Project (MIT Media Lab, USA)

The Open Agriculture Initiative developed controlled-environment "food computers", open-source hardware that lets users test plant growth variables.

By sharing data globally, participants collectively refine sustainable urban food systems.

This project reflects the spirit of permaculture: distributed learning and iterative adaptation.

Case Study: Barcelona's "Fab City" Model

Barcelona's Fab City vision aims to produce everything the city consumes locally by 2050, using digital fabrication, local materials, and circular design.

Makerspaces (Fab Labs) empower citizens to create and repair products, reducing imports and waste. It is a practical fusion of technology, craftsmanship, and ecological responsibility, a blueprint for digital permaculture economies.

Design Exercise

Create a Digital Ecosystem Map:

1. Choose a project, a garden, block, or entire neighborhood.
2. Identify the systems involved: water, food, waste, and energy.
3. Use a simple digital tool (Google My Maps, Miro, or QGIS) to visualize connections.
4. Add data layers, sunlight, rainfall, community members, and potential collaborations.
5. Share it with your local group to spark co-design and innovation.

Digital tools become truly regenerative when they build relationships, not dependencies.

The Ethics of Tech in Permaculture

As with any tool, the intention behind technology determines its impact.

Permaculture asks one essential question: *Does this serve life?*

When the answer is yes, technology becomes sacred, a channel for awareness and balance.

When misused, it becomes another form of extraction.

Our task as regenerative designers is to ensure that every circuit, every sensor, every byte of data supports the healing of the biosphere.

Toward a Symbiotic Future

The future of cities will not be digital *or* natural, it will be symbiotic.

Imagine smart forests that self-report biodiversity.

Buildings that adapt to seasons like trees.

Neighborhoods where data flows like water, nourishing understanding, not consumption.

Technology guided by permaculture doesn't mechanize life, it illuminates its intelligence.

In such cities, wisdom and innovation walk hand in hand, shaping environments that are not only efficient, but deeply alive.

Case Studies around the World

Seeing the Pattern of Regeneration

Across continents, a quiet revolution is transforming how we build, feed, and govern ourselves.

Urban permaculture has moved beyond backyards and gardens into the realms of policy, architecture, business, and education. Each case study here showcases how different cultures have adapted the same ethics, Earth Care, People Care, Fair Share, to their unique bioregional contexts.

Together, they illustrate one truth: there is no single model for regeneration, only a shared pattern of respect for life.

1. Auroville, India – The City of Human Unity

Background

Founded in 1968 in Tamil Nadu, Auroville was envisioned as an international city dedicated to human and ecological evolution.

What Makes It Unique

- Built on 20 square kilometers of once-barren land.
- Over decades, residents have reforested thousands of acres, restored soil fertility, and built energy self-sufficiency.
- The community uses permaculture design in food production, water harvesting, and housing.

Key Systems

- **Water harvesting tanks** and percolation ponds manage seasonal rains.
- **Community-supported farms** provide fresh food for residents.
- **Eco-housing clusters** built with stabilized earth blocks and solar power.

Lesson

Auroville shows that when community and ecology align, a desert can become a forest, and a utopia can become a functioning system.

2. Curitiba, Brazil – Urban Ecology in Action

Overview

Curitiba has been hailed as one of the world's most sustainable cities for decades.

Innovations

- **Integrated public transport** system reducing car dependency.

- **Parks and flood control:** green zones designed as natural water buffers.
- **"Garbage That Is Not Garbage" program:** residents exchange recyclables for food or bus tickets.

Permaculture Principles in Practice

- "Catch and store energy" through resource exchange.
- "Use edges and value the marginal", converting floodplains into biodiversity corridors.
- "Design from patterns to details", seeing city design as a dynamic ecosystem.

Curitiba proves that even large cities can operate like ecological organisms, with efficiency rooted in empathy.

3. Totnes, England – The Birthplace of the Transition Movement

Background

The Transition Town model began in Totnes in 2005 as a response to climate change and peak oil.

Key Elements

- Local energy and food projects.
- Community currency (the "Totnes Pound") supporting local businesses.
- Educational workshops for resilience, from repair cafés to seed swaps.

Impact

Transition Towns have now spread to over 1,000 communities worldwide, showing that citizen-led action can rebuild resilience faster than top-down policy ever could.

4. The EcoCity of Hammarby Sjöstad, Sweden

Background

Hammarby Sjöstad in Stockholm began as a brownfield redevelopment project in the 1990s and has become a model of integrated sustainability.

Design Features

- Closed-loop water, waste, and energy systems.
- Biogas extracted from organic waste powers public transport.
- District heating from renewable sources.
- Green roofs and waterfront parks filter rainwater.

Key Principle

"Integrate rather than segregate", every output feeds another system. It is **permaculture at the scale of a city**, backed by Scandinavian precision and long-term policy vision.

5. Songdo, South Korea – The Smart Regenerative City

Overview

Built from scratch near Incheon, Songdo is one of the world's first "smart cities," integrating green technology and efficient planning.

Innovations

- Pneumatic waste collection system reduces landfill dependence.
- 40% of land dedicated to green space and waterways.
- Real-time energy management through IoT systems.

Challenge and Potential

Though initially tech-driven, Songdo's evolution toward human-centered design shows that digital intelligence must be balanced with ecological empathy.

Permaculture thinking is now being introduced into its community programs and public landscapes.

6. Vancouver Island, Canada – EcoVillage and Coastal Resilience

The Project

O.U.R. Ecovillage on Vancouver Island integrates permaculture with education, housing, and land stewardship.

Features

- Food forests and agroforestry systems.
- Natural building workshops using cob and straw bale.
- Cooperative land trust ensuring permanent protection.

Lesson

Ecovillages like this one demonstrate how **education, community, and economy** can operate as one regenerative cycle, teaching thousands how to rebuild from the ground up.

7. Freetown, Sierra Leone – Urban Forestry for Climate Resilience

Background

Facing deforestation, floods, and landslides, Freetown launched the **#FreetownTheTreeTown** campaign.

Outcomes

- Over 1 million trees planted since 2020.
- Citizens track and monitor trees using a digital app, merging ecology with civic tech.
- Youth employment in nursery and maintenance programs.

Impact

This initiative embodies *People Care* and *Earth Care* simultaneously, turning environmental restoration into a pathway for dignity and employment.

8. Medellín, Colombia – Social Urbanism

Transformation

Once infamous for crime and inequality, Medellín reimagined itself through **inclusive design and public participation.**

Initiatives

- Cable cars and escalators connecting low-income hillside neighborhoods.
- Public parks and libraries in marginalized zones.
- Urban greening and reforestation in steep areas prone to erosion.

Principle

"Design from patterns to details", regeneration begins with accessibility and trust.

Medellín shows how social inclusion is environmental design.

9. Shenzhen, China – Circular Manufacturing and Green Technology

Overview

Shenzhen, once a fishing village, is now a global innovation hub integrating circular economy principles.

Key Initiatives

- Industrial symbiosis: waste from one factory becomes input for another.
- Urban rooftop solar grids and vertical farms.
- Public green corridors linking neighborhoods.

Lesson

When economic power meets ecological ethics, cities can pivot from exploitation to production with regeneration.

10. Cape Town, South Africa – Water Resilience and Community Permaculture

Background

After experiencing near "Day Zero", when taps almost ran dry, Cape Town embraced community-level permaculture and water conservation.

Responses

- Rainwater catchment and greywater recycling in homes.
- Community gardens teaching drought-tolerant agriculture.
- City-level education on water ethics and citizen stewardship.

Outcome

The city averted total collapse, not through technology alone, but through collective behavioral redesign.

Patterns Emerging from Global Practice

Across all continents, successful regenerative cities share five constants:

1. **Integration:** Systems are designed to work together, not in isolation.
2. **Participation:** Citizens co-create, rather than consume solutions.
3. **Localization:** Solutions match cultural and climatic context.
4. **Education:** Regeneration is sustained through learning, not laws alone.

5. **Feedback:** Monitoring and adaptation replace rigid master plans.

The result? Cities that behave more like ecosystems, self-regulating, self-healing, and self-evolving.

Design Exercise

Analyze a Case Study:

1. Select one of the cities above.
2. Identify which permaculture principles it exemplifies.
3. Note what challenges remain unsolved.
4. Imagine an improvement using a regenerative design approach.
5. Share or present your findings, because transformation multiplies through storytelling.

By studying success stories, we plant new seeds in our own bioregions.

The Global Weave of Regeneration

These case studies are more than isolated examples, they are nodes in a planetary network of care.

When viewed together, they reveal humanity's collective evolution from extraction to participation.

From the rice terraces of Asia to the rooftops of Europe, from African reforestation to Latin American cooperatives, the same pattern emerges: life seeks balance, and when design aligns with that law, thriving becomes inevitable.

The future is not built; it is grown, one garden, one community, one conscious city at a time.

The Future of Urban Permaculture: Regenerative, Resilient, and Thriving Cities

A Vision beyond Sustainability

"**Sustainability**" once meant keeping things as they are. But the truth is, what we have built is not worth sustaining.

We need cities that **heal**, not just survive.

That generate more life than they consume.

That feed soil, water, and spirit alike.

The future of urban permaculture is not about sustaining an old world, it is about designing the next one: regenerative, inclusive, and alive.

From Human Habitat to Living Ecosystem

The city of the future will not stand *apart* from nature.

It will function as nature, an integrated web of relationships exchanging energy, matter, and meaning.

Imagine:

- Buildings that photosynthesize.
- Streets that absorb stormwater and filter air.
- Food grown in vertical forests and hydroponic corridors.

- Neighborhoods where waste does not exist, because everything is reused, reimagined, or reborn.

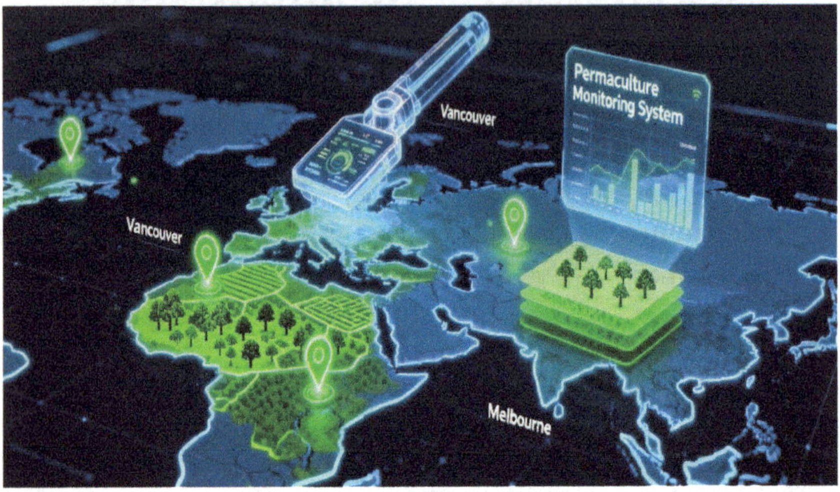

In these living cities, architecture becomes habitat, and governance becomes ecology. Every decision, from transportation to technology, regenerates, rather than extracts.

Principles of the Regenerative City

1. **Circularity:** All materials and energy loop endlessly, nothing becomes waste.
2. **Biophilia:** Every space strengthens our emotional connection to nature.
3. **Resilience:** Cities adapt like ecosystems, diverse, decentralized, and self-regulating.
4. **Equity:** Regeneration is not just environmental; it is social and economic justice.
5. **Feedback:** Data and observation ensure continuous learning and adaptation.

6. **Beauty:** Design uplifts the human spirit, aesthetics as nourishment.

The regenerative city is a living teacher, evolving through observation, creativity, and care.

The Role of Regenerative Design in Urban Transformation

Regenerative design asks not, *"How can we minimize harm?"*, but *"How can we maximize healing?"*

1. Ecological Restoration

Urban rewilding restores biodiversity: pollinator corridors, wetlands, and green roofs reconnect fragmented ecosystems.

2. Social Regeneration

Housing, education, and economy are redesigned around participation and shared value.

When people feel ownership, stewardship naturally follows.

3. Cultural Regeneration

Local art, indigenous wisdom, and community rituals become part of city identity, nurturing belonging in an increasingly digital world.

The city evolves not through demolition, but through re-inhabitation, healing both place and people.

The New Urban Citizen

The future city depends on a new kind of citizen, the ecological human. Someone who sees their neighborhood as a living organism.

Who composts not out of guilt, but reverence.

Who measures prosperity in well-being, not accumulation.

The regenerative citizen:

- Participates in local food and energy networks.
- Designs or maintains a community garden.
- Advocates for fair and transparent governance.
- Educates others, not through theory, but example.

Transformation begins not in city hall, but in the home garden, the shared kitchen, the neighborhood meeting.

Resilience in a Time of Uncertainty

Climate disruption, social polarization, and economic instability are defining our era. But permaculture teaches that crises are **compost**, raw material for transformation.

Resilient cities are built not by avoiding disturbance, but by learning to dance with it.

When systems are flexible, diverse, and cooperative, they adapt rather than collapse.

Resilience means:

- Distributed food and energy systems.
- Localized economies that don't depend on global supply chains.
- Social safety nets woven through mutual aid, not bureaucracy.
- Emotional resilience cultivated through art, mindfulness, and community.

In the regenerative city, resilience is not an emergency response, it is a way of life.

The Regenerative Economy

The future urban economy will no longer measure success by GDP, but by Gross Ecosystem Happiness.

It will prioritize long-term well-being over short-term gain.

Key shifts:

- **Circular Production:** Repair, reuse, and sharing replace disposal.
- **Localized Trade:** Neighborhood markets thrive through digital and physical exchange.
- **Regenerative Finance:** Impact investment and cooperative ownership structures circulate wealth.
- **Eco-literacy in Business:** Every entrepreneur becomes a steward of planetary health.

The regenerative economy operates like a forest, every transaction feeds the system that sustains it.

Education for Regenerative Futures

True transformation begins with learning.

Schools, universities, and community centers will become laboratories for ecological literacy, teaching observation, systems thinking, and collaboration.

Imagine a new generation of professionals:

- Architects trained to mimic forest dynamics.
- Engineers designing like rivers flow.
- Artists crafting with living materials.
- Economists calculating the cost of pollution as loss of life, not line items.

Education will no longer prepare people to consume, it will prepare them to **coexist**.

Technology as Regenerative Intelligence

When guided by permaculture ethics, technology amplifies nature's intelligence rather than replacing it.

Cities will use biofeedback systems, AI for ecosystem management, and digital twins to monitor and balance urban metabolism.

Smart grids will mimic mycelium networks, decentralized, self-correcting, and adaptive.

Data will not belong to corporations, but to communities, used to heal, not to exploit.

Technology and biology will merge not in dominance, but in dialogue.

Governance as Collective Stewardship

In regenerative cities, governance shifts from authority to stewardship.
Citizens co-create policies through assemblies, digital platforms, and participatory budgeting.

Government becomes a facilitator of regeneration, not a regulator of compliance.

Metrics will include:

- Soil health and biodiversity indexes.
- Access to nature per resident.
- Community well-being indicators.
- Climate adaptation readiness.

Leadership becomes less about power and more about care, listening, and co-design.

A Global Network of Regenerative Cities

Already, cities are forming international alliances to share solutions:

- **C40 Cities Climate Leadership Group** advancing low-carbon innovation.
- **ICLEI** promoting local sustainability governance.

- **Regenerative City Network** connecting eco-designers and policymakers.
- **Transition Towns** linking community resilience projects worldwide.

In time, these networks will form a **planetary permaculture web**, each city a node exchanging energy, wisdom, and creativity. When cities cooperate like ecosystems, humanity itself becomes regenerative.

Design Exercise

Visioning the Regenerative City of 2050:

1. Envision your city in 25 years if permaculture principles were fully implemented.
2. Describe how people live, move, and work.
3. Sketch what public spaces look like, food forests, rooftop farms, biophilic buildings.
4. Imagine the emotional tone, calm, connected, creative.
5. Ask yourself: *What one action today moves us closer to that world?*

Vision is the seed of design, and design is the first act of regeneration.

The Ethical Imperative

The final lesson of urban permaculture is simple yet profound: We are not separate from the systems we design. Our cities are mirrors of our values.

To regenerate cities, we must first regenerate consciousness, shifting from dominance to partnership, from scarcity to abundance, from fear to care.

The city of the future begins with a single question:

What kind of ancestor do we wish to be?

From Sustainability to Thrivability

When regeneration becomes culture, and design becomes compassion, cities will no longer be engines of extraction, they will be gardens of belonging.

Streets will hum with pollinators. Waterways will sparkle again.

Children will know where their food comes from.

And every act, from cooking to policymaking, will express reverence for life.

The regenerative city is not a dream, it is a design we are already sketching together.

The blueprint is written not in blue ink, but in green living systems, evolving wherever care takes root.

Conclusion

The City as a Living Promise

Every city is a story. A story written in stone, steel, and soil, shaped by what we value, and by how we live together.

For centuries, humanity built cities as fortresses against nature, walls of progress meant to separate us from the cycles that sustain us. Now, the task before us is the opposite: to **reunite**. To design cities not as machines, but as **living organisms**, breathing, adapting, and evolving with the Earth itself.

Urban permaculture invites us to become authors of a new kind of story: one where care replaces consumption, where cooperation outgrows competition, and where regeneration becomes civilization's defining art.

What We've Learned

Across this journey, we explored how permaculture reshapes not only gardens, but governance, economics, education, and consciousness itself.

We learned that:

- Nature is not a resource, it is our teacher.
- Design is not decoration, it is ethics in motion.
- Food systems, energy, and architecture can coexist in harmony.
- Policy and community are most powerful when they collaborate, not compete.
- Technology is most intelligent when it listens to life, not replaces it.
- Healing the planet begins with healing our relationship to it.

Permaculture is not a style of gardening. It is a philosophy of participation, a reminder that humans belong to the Earth's metabolism, not above it.

From Knowledge to Practice

Understanding these principles is only the first step.

Regeneration begins when ideas take root in daily behavior, when compost replaces waste, when dialogue replaces division, when empathy guides design.

Every person, every household, every community can act as a node of regeneration:

- Start a small garden, even on a balcony.
- Support local growers and cooperatives.
- Plant trees where there were none.
- Collect rainwater, reuse greywater.
- Teach others what you learn.

No act is too small. In ecology, the smallest system often triggers the largest change.

The Power of Community

The regenerative city is not built by corporations or governments alone, it is built by relationships.

By people who care for each other and for place.

When citizens take ownership of their neighborhoods, when decisions are made through dialogue, when resources circulate locally, cities begin to thrive as ecosystems.

Permaculture turns urban life from survival into **belonging**, each person a steward, each space a story of care.

Healing as Design

Ultimately, permaculture is the design of healing.

Healing soil, water, air, but also the emotional, cultural, and spiritual landscapes we inhabit.

Cities mirror our inner world.

If we are fragmented, they are chaotic.

If we are compassionate, they become coherent.

Urban permaculture teaches that regeneration is not just an external process, it is an inner practice of attention, patience, and love.

When we learn to see the city as alive, as something we can nurture, not exploit, every act of design becomes sacred.

A Call to Future Designers

You, the reader, are now part of this movement. Whether you are an architect, teacher, artist, policymaker, or gardener, your work matters.

The future belongs to those who design for life.

Let your projects heal as they create.

Let your decisions serve generations unseen.

Let your imagination become infrastructure for hope.

The tools are already in our hands.

What remains is the courage to use them with love.

The Great Turning

Humanity stands at a crossroads: between extraction and regeneration, isolation and community, collapse and renewal.

Permaculture offers a map, not of control, but of cooperation.

It shows us that we can reimagine cities not as burdens on the planet, but as gardens of abundance, each one a seed of global restoration.

The Great Turning is already underway, in gardens, rooftops, policy rooms, and classrooms.

Every compost pile, every shared meal, every cooperative decision brings us closer to the world we long for.

The Closing Circle

As we come to the end of this book, remember: there is no "end" in permaculture.

Only cycles. Only beginnings disguised as endings.

Each chapter of this book mirrors a pattern in nature, observation, design, integration, feedback, and evolution.

And like a forest, the story continues through you.

Because the true city of the future is not somewhere else, it is the one we are planting today.

A Final Reflection

"To build a regenerative city is to remember that we are not apart from the Earth, we are the Earth, designing herself back into balance."

Let this book be not a manual, but an invitation. Go outside. Observe. Plant something. Talk to your neighbors. Listen to the wind between buildings, the water under streets, the roots that still find a way to rise.

There, in that listening, you will find the blueprint, not just for the city, but for life itself.

Appendix

A. Key Frameworks and Methodologies

1. The Permaculture Design Process

A simple framework that can be applied to any urban project, from a balcony garden to an entire district redevelopment.

1. **Observation**: Study the site's patterns: sun, shade, water, people, and energy flows.
2. **Analysis**: Identify constraints, opportunities, and relationships between systems.
3. **Design**: Apply permaculture principles to integrate ecology, function, and beauty.
4. **Implementation**: Start small, work in phases, and use available local resources.
5. **Maintenance**: Monitor and adjust through feedback loops.
6. **Reflection**: Evaluate results, document lessons, and evolve the design.

This cycle mirrors natural succession, learning and adaptation are continuous.

2. The 12 Principles of Permaculture (by David Holmgren)

1. Observe and interact.
2. Catch and store energy.
3. Obtain a yield.

4. Apply self-regulation and accept feedback.
5. Use and value renewable resources and services.
6. Produce no waste.
7. Design from patterns to details.
8. Integrate rather than segregate.
9. Use small and slow solutions.
10. Use and value diversity.
11. Use edges and value the marginal.
12. Creatively use and respond to change.

These principles are universal, adaptable to landscapes, organizations, and even personal development.

3. The Three Ethics of Permaculture

1. **Earth Care**: Protect and regenerate the natural systems that sustain life.
2. **People Care**: Support well-being, connection, and community resilience.
3. **Fair Share (or Future Care)**: Limit consumption, share surplus, and ensure justice across generations.

All regenerative systems must align with these ethics, they are the compass that ensures right direction over mere efficiency.

B. Recommended Reading and References

Foundational Works

- *Permaculture: Principles and Pathways Beyond Sustainability*, David Holmgren
- *Permaculture: A Designers' Manual*, Bill Mollison
- *Gaia's Garden: A Guide to Home-Scale Permaculture*, Toby Hemenway
- *The Transition Handbook*, Rob Hopkins
- *Cradle to Cradle: Remaking the Way We Make Things*, William McDonough & Michael Braungart
- *The Nature of Order*, Christopher Alexander
- *Biophilic Design: The Theory, Science, and Practice of Bringing Buildings to Life*, Stephen Kellert & Judith Heerwagen

Urban and Social Focus

- *Designing Regenerative Cultures*, Daniel Christian Wahl
- *The Regenerative Business*, Carol Sanford
- *Doughnut Economics*, Kate Raworth
- *The Ecological City*, Hagan & Houchin
- *Resilient Cities*, Peter Newman, Timothy Beatley, Heather Boyer

Case Studies and Global Insights

- *The Transition Companion*, Rob Hopkins
- *The EcoCities Reader*, Richard Register

- *Greening the Urban Economy*, OECD
- *Regenerative Urban Design and Ecosystem Biomimicry*, Maibritt Pedersen Zari

C. Online Courses and Training

Permaculture Design Certification (PDC) Programs

- **Permaculture Research Institute (Australia):** www.permaculturenews.org
- **Permaculture Institute of North America (PINA):** www.pina.institute
- **Gaia University (Global):** www.gaiauniversity.org
- **Urban Permaculture Institute (San Francisco):** www.upisf.com
- **Permaculture Association (UK):** www.permaculture.org.uk

Most of these offer hybrid options, combining online learning with local design practicums.

D. Global Networks and Organizations

Network	Focus	Website
Transition Network	Community-led resilience & localization	www.transitionnetwork.org
Regenerative Communities Network (Capital Institute)	Systems transformation & bioregional economies	www.capitalinstitute.org
ICLEI – Local Governments	Municipal sustainability	www.iclei.org

for Sustainability	collaboration	
C40 Cities	Global network of cities addressing climate change	www.c40.org
EcoVillage Network (GEN)	Ecovillage development and social permaculture	www.ecovillage.org
Urban Food Policy Pact	Global urban food system cooperation	www.milanurbanfoodpolicypact.org

The networks help practitioners exchange ideas, access funding, and amplify their work.

E. Tools and Platforms for Urban Designers

Tool	Purpose	Notes
QGIS	GIS mapping and spatial analysis	Free and open source
SketchUp	3D landscape and architectural design	Ideal for visualizing small sites
Trello / Notion / Asana	Project coordination	Excellent for community collaboration
OpenStreetMap	Mapping community assets	Can map gardens, compost hubs, and water flows
iNaturalist	Citizen science & biodiversity tracking	Encourages community participation
Google Earth Pro	Site observation and data visualization	Perfect for macro-scale analysis

F. Permaculture Design Templates

1. **Urban Site Survey Checklist**

- Sun and shade mapping
- Wind direction and microclimate zones
- Water flow and drainage
- Existing vegetation
- Soil type and structure
- Access points and human movement patterns

2. **Design Elements Integration Map**

- Food systems (gardens, vertical farms, edible landscaping)
- Energy systems (solar, passive design, microgrids)
- Water systems (harvesting, filtration, reuse)
- Waste cycles (compost, recycling, zero-waste streams)
- Social systems (community centers, co-ops, learning hubs)

These templates ensure no element exists in isolation, mirroring nature's interconnectedness.

G. Funding and Support Opportunities

- **UN-Habitat Urban Resilience Fund**: for city-level sustainability projects.
- **European Green Capital & Green Leaf Awards**: support municipal innovation.

- **Local crowd-investing platforms** (e.g., GoFundMe, Open Collective) for community gardens.
- **Impact Investment Funds**: such as RSF Social Finance or GreenBank.
- **Corporate ESG Partnerships**: many companies now fund urban greening as carbon offset initiatives.

Tip: Always link your project's outcomes to social and ecological benefits, funders respond to measurable impact.

H. Glossary of Key Terms

Term	Definition
Agroecology	The integration of ecological principles into agricultural systems.
Biophilic Design	Architecture that connects people with nature.
Circular Economy	An economy where resources are reused and waste eliminated.
Ecovillage	A community designed for ecological and social sustainability.
Guild	A group of complementary plants or elements that support each other's function.
Resilience	The capacity of a system to recover and adapt to change.
Social Permaculture	Application of permaculture principles to human systems and relationships.
Transition Town	A community mobilized toward local resilience and low-carbon living.

I. Reflection Journal Prompts

1. What is one small space in your life or neighborhood that could be redesigned for regeneration?

2. Which permaculture principle do you naturally practice, and which challenges you most?
3. How can your professional or creative skills serve a regenerative goal?
4. Who in your community shares your vision, and how can you collaborate?

What legacy do you want to leave in your city?

Reflection transforms awareness into purpose.

Purpose transforms purpose into action.

J. Final Words of Encouragement

Urban permaculture is not a movement for experts, it is a movement for everyone.

Every balcony garden, compost bin, seed swap, and rooftop hive matters.

Each small system, each cooperative, each conversation adds up to a planetary transformation already underway.

You are now part of that story, the next generation of city makers, designing with nature instead of against it.

The journey does not end here.

This book was your seed packet.

What you grow from it, that's your forest.

"The city that learns to care for its soil will never run out of hope."

Made in the USA
Monee, IL
03 November 2026

67382299R00115